Practice

Eureka Math®
Grade 3 Fluency
Modules 5–7

W9-AVJ-485

Published by Great Minds®.

Copyright © 2015 Great Minds®. No part of this work may be reproduced, sold, or commercialized, in whole or in part, without written permission from Great Minds®. Noncommercial use is licensed pursuant to a Creative Commons Attribution-NonCommercial-ShareAlike 4.0 license; for more information, go to http://greatminds.org/copyright. *Great Minds* and *Eureka Math* are registered trademarks of Great Minds®.

Printed in the U.S.A.
This book may be purchased from the publisher at eureka-math.org.
3 4 5 6 7 8 9 10   CCR   24 23 22

ISBN 978-1-64054-623-3

G3-M5-M7-P/F-04.2018

Learn ◆ Practice ◆ Succeed

Eureka Math® student materials for *A Story of Units®* (K–5) are available in the *Learn, Practice, Succeed* trio. This series supports differentiation and remediation while keeping student materials organized and accessible. Educators will find that the *Learn, Practice,* and *Succeed* series also offers coherent—and therefore, more effective—resources for Response to Intervention (RTI), extra practice, and summer learning.

Learn

Eureka Math Learn serves as a student's in-class companion where they show their thinking, share what they know, and watch their knowledge build every day. *Learn* assembles the daily classwork—Application Problems, Exit Tickets, Problem Sets, templates—in an easily stored and navigated volume.

Practice

Each *Eureka Math* lesson begins with a series of energetic, joyous fluency activities, including those found in *Eureka Math Practice.* Students who are fluent in their math facts can master more material more deeply. With *Practice,* students build competence in newly acquired skills and reinforce previous learning in preparation for the next lesson.

Together, *Learn* and *Practice* provide all the print materials students will use for their core math instruction.

Succeed

Eureka Math Succeed enables students to work individually toward mastery. These additional problem sets align lesson by lesson with classroom instruction, making them ideal for use as homework or extra practice. Each problem set is accompanied by a Homework Helper, a set of worked examples that illustrate how to solve similar problems.

Teachers and tutors can use *Succeed* books from prior grade levels as curriculum-consistent tools for filling gaps in foundational knowledge. Students will thrive and progress more quickly as familiar models facilitate connections to their current grade-level content.

Students, families, and educators:

Thank you for being part of the *Eureka Math®* community, where we celebrate the joy, wonder, and thrill of mathematics. One of the most obvious ways we display our excitement is through the fluency activities provided in *Eureka Math Practice*.

What is fluency in mathematics?

You may think of *fluency* as associated with the language arts, where it refers to speaking and writing with ease. In prekindergarten through grade 5, the *Eureka Math* curriculum contains multiple daily opportunities to build fluency *in mathematics*. Each is designed with the same notion—growing every student's ability to use mathematics *with ease*. Fluency experiences are generally fast-paced and energetic, celebrating improvement and focusing on recognizing patterns and connections within the material. They are not intended to be graded.

Eureka Math fluency activities provide differentiated practice through a variety of formats—some are conducted orally, some use manipulatives, others use a personal whiteboard, and still others use a handout and paper-and-pencil format. *Eureka Math Practice* provides each student with the printed fluency exercises for his or her grade level.

What is a Sprint?

Many printed fluency activities utilize the format we call a Sprint. These exercises build speed and accuracy with already acquired skills. Used when students are nearing optimum proficiency, Sprints leverage tempo to build a low-stakes adrenaline boost that increases memory and recall. Their intentional design makes Sprints inherently differentiated; the problems build from simple to complex, with the first quadrant of problems being the simplest and each subsequent quadrant adding complexity. Further, intentional patterns within the sequence of problems engage students' higher order thinking skills.

The suggested format for delivering a Sprint calls for students to do two consecutive Sprints (labeled A and B) on the same skill, each timed at one minute. Students pause between Sprints to articulate the patterns they noticed as they worked the first Sprint. Noticing the patterns often provides a natural boost to their performance on the second Sprint.

Sprints can be conducted with an untimed protocol as well. The untimed protocol is highly recommended when students are still building confidence with the level of complexity of the first quadrant of problems. Once all students are prepared for success on the Sprint, the work of improving speed and accuracy with the energy of a timed protocol is often welcome and invigorating.

Where can I find other fluency activities?

The *Eureka Math Teacher Edition* guides educators in the delivery of all fluency activities for each lesson, including those that do not require print materials. Additionally, the *Eureka Digital Suite* provides access to the fluency activities for all grade levels, searchable by standard or lesson.

Best wishes for a year filled with aha moments!

Jill Diniz

Jill Diniz
Director of Mathematics
Great Minds

Contents

Module 5

Module 6

Module 7

© 2015 Great Minds® eureka-math.org

Grade 3
Module 5

A

Number Correct: _____

Multiply with Six

1.	$1 \times 6 =$	
2.	$6 \times 1 =$	
3.	$2 \times 6 =$	
4.	$6 \times 2 =$	
5.	$3 \times 6 =$	
6.	$6 \times 3 =$	
7.	$4 \times 6 =$	
8.	$6 \times 4 =$	
9.	$5 \times 6 =$	
10.	$6 \times 5 =$	
11.	$6 \times 6 =$	
12.	$7 \times 6 =$	
13.	$6 \times 7 =$	
14.	$8 \times 6 =$	
15.	$6 \times 8 =$	
16.	$9 \times 6 =$	
17.	$6 \times 9 =$	
18.	$10 \times 6 =$	
19.	$6 \times 10 =$	
20.	$6 \times 3 =$	
21.	$1 \times 6 =$	
22.	$2 \times 6 =$	

23.	$10 \times 6 =$	
24.	$9 \times 6 =$	
25.	$4 \times 6 =$	
26.	$8 \times 6 =$	
27.	$3 \times 6 =$	
28.	$7 \times 6 =$	
29.	$6 \times 6 =$	
30.	$6 \times 10 =$	
31.	$6 \times 5 =$	
32.	$6 \times 4 =$	
33.	$6 \times 1 =$	
34.	$6 \times 9 =$	
35.	$6 \times 6 =$	
36.	$6 \times 3 =$	
37.	$6 \times 2 =$	
38.	$6 \times 7 =$	
39.	$6 \times 8 =$	
40.	$11 \times 6 =$	
41.	$6 \times 11 =$	
42.	$12 \times 6 =$	
43.	$6 \times 12 =$	
44.	$13 \times 6 =$	

EUREKA MATH

Lesson 3: Specify and partition a whole into equal parts, identifying and counting unit fractions by drawing pictorial area models.

3

© 2015 Great Minds®. eureka-math.org

B

Number Correct: _____

Multiply with Six

Improvement: _____

1.	6 × 1 =	
2.	1 × 6 =	
3.	6 × 2 =	
4.	2 × 6 =	
5.	6 × 3 =	
6.	3 × 6 =	
7.	6 × 4 =	
8.	4 × 6 =	
9.	6 × 5 =	
10.	5 × 6 =	
11.	6 × 6 =	
12.	6 × 7 =	
13.	7 × 6 =	
14.	6 × 8 =	
15.	8 × 6 =	
16.	6 × 9 =	
17.	9 × 6 =	
18.	6 × 10 =	
19.	10 × 6 =	
20.	1 × 6 =	
21.	10 × 6 =	
22.	2 × 6 =	

23.	9 × 6 =	
24.	3 × 6 =	
25.	8 × 6 =	
26.	4 × 6 =	
27.	7 × 6 =	
28.	5 × 6 =	
29.	6 × 6 =	
30.	6 × 5 =	
31.	6 × 10 =	
32.	6 × 1 =	
33.	6 × 6 =	
34.	6 × 4 =	
35.	6 × 9 =	
36.	6 × 2 =	
37.	6 × 7 =	
38.	6 × 3 =	
39.	6 × 8 =	
40.	11 × 6 =	
41.	6 × 11 =	
42.	12 × 6 =	
43.	6 × 12 =	
44.	13 × 6 =	

Lesson 3:   Specify and partition a whole into equal parts, identifying and counting unit fractions by drawing pictorial area models.

5

© 2015 Great Minds®. eureka-math.org

A

Number Correct: _____

Multiply and Divide by Six

1.	2 × 6 =	
2.	3 × 6 =	
3.	4 × 6 =	
4.	5 × 6 =	
5.	1 × 6 =	
6.	12 ÷ 6 =	
7.	18 ÷ 6 =	
8.	30 ÷ 6 =	
9.	6 ÷ 6 =	
10.	24 ÷ 6 =	
11.	6 × 6 =	
12.	7 × 6 =	
13.	8 × 6 =	
14.	9 × 6 =	
15.	10 × 6 =	
16.	48 ÷ 6 =	
17.	42 ÷ 6 =	
18.	54 ÷ 6 =	
19.	36 ÷ 6 =	
20.	60 ÷ 6 =	
21.	___ × 6 = 30	
22.	___ × 6 = 6	

23.	___ × 6 = 60	
24.	___ × 6 = 12	
25.	___ × 6 = 18	
26.	60 ÷ 6 =	
27.	30 ÷ 6 =	
28.	6 ÷ 6 =	
29.	12 ÷ 6 =	
30.	18 ÷ 6 =	
31.	___ × 6 = 36	
32.	___ × 6 = 42	
33.	___ × 6 = 54	
34.	___ × 6 = 48	
35.	42 ÷ 6 =	
36.	54 ÷ 6 =	
37.	36 ÷ 6 =	
38.	48 ÷ 6 =	
39.	11 × 6 =	
40.	66 ÷ 6 =	
41.	12 × 6 =	
42.	72 ÷ 6 =	
43.	14 × 6 =	
44.	84 ÷ 6 =	

© 2015 Great Minds®. eureka-math.org

B

Number Correct: _____

Multiply and Divide by Six

Improvement: _____

1.	1 × 6 =	
2.	2 × 6 =	
3.	3 × 6 =	
4.	4 × 6 =	
5.	5 × 6 =	
6.	18 ÷ 6 =	
7.	12 ÷ 6 =	
8.	24 ÷ 6 =	
9.	6 ÷ 6 =	
10.	30 ÷ 6 =	
11.	10 × 6 =	
12.	6 × 6 =	
13.	7 × 6 =	
14.	8 × 6 =	
15.	9 × 6 =	
16.	42 ÷ 6 =	
17.	36 ÷ 6 =	
18.	48 ÷ 6 =	
19.	60 ÷ 6 =	
20.	54 ÷ 6 =	
21.	____ × 6 = 6	
22.	____ × 6 = 30	

23.	____ × 6 = 12	
24.	____ × 6 = 60	
25.	____ × 6 = 18	
26.	12 ÷ 6 =	
27.	6 ÷ 6 =	
28.	60 ÷ 6 =	
29.	30 ÷ 6 =	
30.	18 ÷ 6 =	
31.	____ × 6 = 18	
32.	____ × 6 = 24	
33.	____ × 6 = 54	
34.	____ × 6 = 42	
35.	48 ÷ 6 =	
36.	54 ÷ 6 =	
37.	36 ÷ 6 =	
38.	42 ÷ 6 =	
39.	11 × 6 =	
40.	66 ÷ 6 =	
41.	12 × 6 =	
42.	72 ÷ 6 =	
43.	13 × 6 =	
44.	78 ÷ 6 =	

© 2015 Great Minds®. eureka-math.org

A

Number Correct: _____

Multiply with Seven

1.	1 × 7 =	
2.	7 × 1 =	
3.	2 × 7 =	
4.	7 × 2 =	
5.	3 × 7 =	
6.	7 × 3 =	
7.	4 × 7 =	
8.	7 × 4 =	
9.	5 × 7 =	
10.	7 × 5 =	
11.	6 × 7 =	
12.	7 × 6 =	
13.	7 × 7 =	
14.	8 × 7 =	
15.	7 × 8 =	
16.	9 × 7 =	
17.	7 × 9 =	
18.	10 × 7 =	
19.	7 × 10 =	
20.	7 × 3 =	
21.	1 × 7 =	
22.	2 × 7 =	

23.	10 × 7 =	
24.	9 × 7 =	
25.	4 × 7 =	
26.	8 × 7 =	
27.	7 × 3 =	
28.	7 × 7 =	
29.	6 × 7 =	
30.	7 × 10 =	
31.	7 × 5 =	
32.	7 × 6 =	
33.	7 × 1 =	
34.	7 × 9 =	
35.	7 × 4 =	
36.	7 × 3 =	
37.	7 × 2 =	
38.	7 × 7 =	
39.	7 × 8 =	
40.	11 × 7 =	
41.	7 × 11 =	
42.	12 × 7 =	
43.	7 × 12 =	
44.	13 × 7 =	

Lesson 6:   Build non-unit fractions less than one whole from unit fractions.

11

© 2015 Great Minds®. eureka-math.org

B

Number Correct: _____

Multiply with Seven

Improvement: _____

1.	7 × 1 =	
2.	1 × 7 =	
3.	7 × 2 =	
4.	2 × 7 =	
5.	7 × 3 =	
6.	3 × 7 =	
7.	7 × 4 =	
8.	4 × 7 =	
9.	7 × 5 =	
10.	5 × 7 =	
11.	7 × 6 =	
12.	6 × 7 =	
13.	7 × 7 =	
14.	7 × 8 =	
15.	8 × 7 =	
16.	7 × 9 =	
17.	9 × 7 =	
18.	7 × 10 =	
19.	10 × 7 =	
20.	1 × 7 =	
21.	10 × 7 =	
22.	2 × 7 =	

23.	9 × 7 =	
24.	3 × 7 =	
25.	8 × 7 =	
26.	4 × 7 =	
27.	7 × 7 =	
28.	5 × 7 =	
29.	6 × 7 =	
30.	7 × 5 =	
31.	7 × 10 =	
32.	7 × 1 =	
33.	7 × 6 =	
34.	7 × 4 =	
35.	7 × 9 =	
36.	7 × 2 =	
37.	7 × 7 =	
38.	7 × 3 =	
39.	7 × 8 =	
40.	11 × 7 =	
41.	7 × 11 =	
42.	12 × 7 =	
43.	7 × 12 =	
44.	13 × 7 =	

Lesson 6:   Build non-unit fractions less than one whole from unit fractions.

13

© 2015 Great Minds®. eureka-math.org

A

Number Correct: _____

Multiply and Divide by Seven

1.	2 × 7 =	
2.	3 × 7 =	
3.	4 × 7 =	
4.	5 × 7 =	
5.	1 × 7 =	
6.	14 ÷ 7 =	
7.	21 ÷ 7 =	
8.	35 ÷ 7 =	
9.	7 ÷ 7 =	
10.	28 ÷ 7 =	
11.	6 × 7 =	
12.	7 × 7 =	
13.	8 × 7 =	
14.	9 × 7 =	
15.	10 × 7 =	
16.	56 ÷ 7 =	
17.	49 ÷ 7 =	
18.	63 ÷ 7 =	
19.	42 ÷ 7 =	
20.	70 ÷ 7 =	
21.	___ × 7 = 35	
22.	___ × 7 = 7	

23.	___ × 7 = 70	
24.	___ × 7 = 14	
25.	___ × 7 = 21	
26.	70 ÷ 7 =	
27.	35 ÷ 7 =	
28.	7 ÷ 7 =	
29.	14 ÷ 7 =	
30.	21 ÷ 7 =	
31.	___ × 7 = 42	
32.	___ × 7 = 49	
33.	___ × 7 = 63	
34.	___ × 7 = 56	
35.	49 ÷ 7 =	
36.	63 ÷ 7 =	
37.	42 ÷ 7 =	
38.	56 ÷ 7 =	
39.	11 × 7 =	
40.	77 ÷ 7 =	
41.	12 × 7 =	
42.	84 ÷ 7 =	
43.	14 × 7 =	
44.	98 ÷ 7 =	

Lesson 7:   Identify and represent shaded and non-shaded parts of one whole as
             fractions.

15

© 2015 Great Minds®. eureka-math.org

B

Number Correct: _____

Multiply and Divide by Seven

Improvement: _____

1.	1 × 7 =	
2.	2 × 7 =	
3.	3 × 7 =	
4.	4 × 7 =	
5.	5 × 7 =	
6.	21 ÷ 7 =	
7.	14 ÷ 7 =	
8.	28 ÷ 7 =	
9.	7 ÷ 7 =	
10.	35 ÷ 7 =	
11.	10 × 7 =	
12.	6 × 7 =	
13.	7 × 7 =	
14.	8 × 7 =	
15.	9 × 7 =	
16.	49 ÷ 7 =	
17.	42 ÷ 7 =	
18.	56 ÷ 7 =	
19.	70 ÷ 7 =	
20.	63 ÷ 7 =	
21.	___ × 7 = 7	
22.	___ × 7 = 35	

23.	___ × 7 = 14	
24.	___ × 7 = 70	
25.	___ × 7 = 21	
26.	14 ÷ 7 =	
27.	7 ÷ 7 =	
28.	70 ÷ 7 =	
29.	35 ÷ 7 =	
30.	21 ÷ 7 =	
31.	___ × 7 = 21	
32.	___ × 7 = 28	
33.	___ × 7 = 63	
34.	___ × 7 = 49	
35.	56 ÷ 7 =	
36.	63 ÷ 7 =	
37.	42 ÷ 7 =	
38.	49 ÷ 7 =	
39.	11 × 7 =	
40.	77 ÷ 7 =	
41.	12 × 7 =	
42.	84 ÷ 7 =	
43.	13 × 7 =	
44.	91 ÷ 7 =	

Lesson 7: Identify and represent shaded and non-shaded parts of one whole as fractions.

© 2015 Great Minds®. eureka-math.org

A

Number Correct: _____

Identify Fractions.

1.		/
2.		/
3.		/
4.		/
5.		/
6.		/
7.		/
8.		/
9.		/
10.		/
11.		/
12.		/
13.		/
14.		/
15.		/
16.		/
17.		/
18.		/
19.		/
20.		/
21.		/
22.		/

23.		/
24.		/
25.		/
26.		/
27.		/
28.		/
29.		/
30.		/
31.		/
32.		/
33.		/
34.		/
35.		/
36.		/
37.		/
38.		/
39.		/
40.		/
41.		/
42.		/
43.		/
44.		/

Lesson 8:   Represent parts of one whole as fractions with number bonds.   19

© 2015 Great Minds®. eureka-math.org

B

Number Correct: _____

Identify Fractions.

Improvement: _____

1.		/
2.		/
3.		/
4.		/
5.		/
6.		/
7.		/
8.		/
9.		/
10.		/
11.		/
12.		/
13.		/
14.		/
15.		/
16.		/
17.		/
18.		/
19.		/
20.		/
21.		/
22.		/

23.		/
24.		/
25.		/
26.		/
27.		/
28.		/
29.		/
30.		/
31.		/
32.		/
33.		/
34.		/
35.		/
36.		/
37.		/
38.		/
39.		/
40.		/
41.		/
42.		/
43.		/
44.		/

Lesson 8: Represent parts of one whole as fractions with number bonds.

21

A

Number Correct: _____

Multiply with Eight

1.	8 × 1 =	
2.	1 × 8 =	
3.	8 × 2 =	
4.	2 × 8 =	
5.	8 × 3 =	
6.	3 × 8 =	
7.	8 × 4 =	
8.	4 × 8 =	
9.	8 × 5 =	
10.	5 × 8 =	
11.	8 × 6 =	
12.	6 × 8 =	
13.	8 × 7 =	
14.	7 × 8 =	
15.	8 × 8 =	
16.	8 × 9 =	
17.	9 × 8 =	
18.	8 × 10 =	
19.	10 × 8 =	
20.	1 × 8 =	
21.	10 × 8 =	
22.	2 × 8 =	

23.	9 × 8 =	
24.	3 × 8 =	
25.	8 × 8 =	
26.	4 × 8 =	
27.	7 × 8 =	
28.	5 × 8 =	
29.	6 × 8 =	
30.	8 × 5 =	
31.	8 × 10 =	
32.	8 × 1 =	
33.	8 × 6 =	
34.	8 × 4 =	
35.	8 × 9 =	
36.	8 × 2 =	
37.	8 × 7 =	
38.	8 × 3 =	
39.	8 × 8 =	
40.	11 × 8 =	
41.	8 × 11 =	
42.	12 × 8 =	
43.	8 × 12 =	
44.	13 × 8 =	

© 2013 Great Minds®. eureka-math.org

B

Number Correct: _____

Multiply with Eight

Improvement: _____

1.	1 × 8 =	
2.	8 × 1 =	
3.	2 × 8 =	
4.	8 × 2 =	
5.	3 × 8 =	
6.	8 × 3 =	
7.	4 × 8 =	
8.	8 × 4 =	
9.	5 × 8 =	
10.	8 × 5 =	
11.	6 × 8 =	
12.	8 × 6 =	
13.	7 × 8 =	
14.	8 × 7 =	
15.	8 × 8 =	
16.	9 × 8 =	
17.	8 × 9 =	
18.	10 × 8 =	
19.	8 × 10 =	
20.	8 × 3 =	
21.	1 × 8 =	
22.	2 × 8 =	

23.	10 × 8 =	
24.	9 × 8 =	
25.	4 × 8 =	
26.	8 × 8 =	
27.	8 × 3 =	
28.	7 × 8 =	
29.	6 × 8 =	
30.	8 × 10 =	
31.	8 × 5 =	
32.	8 × 6 =	
33.	8 × 1 =	
34.	8 × 9 =	
35.	8 × 4 =	
36.	8 × 3 =	
37.	8 × 2 =	
38.	8 × 7 =	
39.	8 × 8 =	
40.	11 × 8 =	
41.	8 × 11 =	
42.	12 × 8 =	
43.	8 × 12 =	
44.	13 × 8 =	

Lesson 9:   Build and write fractions greater than one whole using unit fractions.

25

© 2015 Great Minds®. eureka-math.org

A

Number Correct: _____

Multiply and Divide by Eight

1.	2 × 8 =	
2.	3 × 8 =	
3.	4 × 8 =	
4.	5 × 8 =	
5.	1 × 8 =	
6.	16 ÷ 8 =	
7.	24 ÷ 8 =	
8.	40 ÷ 8 =	
9.	8 ÷ 8 =	
10.	32 ÷ 8 =	
11.	6 × 8 =	
12.	7 × 8 =	
13.	8 × 8 =	
14.	9 × 8 =	
15.	10 × 8 =	
16.	64 ÷ 8 =	
17.	56 ÷ 8 =	
18.	72 ÷ 8 =	
19.	48 ÷ 8 =	
20.	80 ÷ 8 =	
21.	___ × 8 = 40	
22.	___ × 8 = 8	

23.	___ × 8 = 80	
24.	___ × 8 = 16	
25.	___ × 8 = 24	
26.	80 ÷ 8 =	
27.	40 ÷ 8 =	
28.	8 ÷ 8 =	
29.	16 ÷ 8 =	
30.	24 ÷ 8 =	
31.	___ × 8 = 48	
32.	___ × 8 = 56	
33.	___ × 8 = 72	
34.	___ × 8 = 64	
35.	56 ÷ 8 =	
36.	72 ÷ 8 =	
37.	48 ÷ 8 =	
38.	64 ÷ 8 =	
39.	11 × 8 =	
40.	88 ÷ 8 =	
41.	12 × 8 =	
42.	96 ÷ 8 =	
43.	14 × 8 =	
44.	112 ÷ 8 =	

Lesson 10:   Compare unit fractions by reasoning about their size using fraction strips.

27

© 2015 Great Minds®. eureka-math.org

B

Number Correct: _____

Multiply and Divide by Eight

Improvement: _____

1.	1 × 8 =	
2.	2 × 8 =	
3.	3 × 8 =	
4.	4 × 8 =	
5.	5 × 8 =	
6.	24 ÷ 8 =	
7.	16 ÷ 8 =	
8.	32 ÷ 8 =	
9.	8 ÷ 8 =	
10.	40 ÷ 8 =	
11.	10 × 8 =	
12.	6 × 8 =	
13.	7 × 8 =	
14.	8 × 8 =	
15.	9 × 8 =	
16.	56 ÷ 8 =	
17.	48 ÷ 8 =	
18.	64 ÷ 8 =	
19.	80 ÷ 8 =	
20.	72 ÷ 8 =	
21.	___ × 8 = 8	
22.	___ × 8 = 40	

23.	___ × 8 = 16	
24.	___ × 8 = 80	
25.	___ × 8 = 24	
26.	16 ÷ 8 =	
27.	8 ÷ 8 =	
28.	80 ÷ 8 =	
29.	40 ÷ 8 =	
30.	24 ÷ 8 =	
31.	___ × 8 = 24	
32.	___ × 8 = 32	
33.	___ × 8 = 72	
34.	___ × 8 = 56	
35.	64 ÷ 8 =	
36.	72 ÷ 8 =	
37.	48 ÷ 8 =	
38.	56 ÷ 8 =	
39.	11 × 8 =	
40.	88 ÷ 8 =	
41.	12 × 8 =	
42.	96 ÷ 8 =	
43.	13 × 8 =	
44.	104 ÷ 8 =	

Lesson 10: Compare unit fractions by reasoning about their size using fraction strips.

29

© 2015 Great Minds®. eureka-math.org

A

Number Correct: _____

Multiply with Nine

1.	$9 \times 1 =$		23.	$9 \times 9 =$	
2.	$1 \times 9 =$		24.	$3 \times 9 =$	
3.	$9 \times 2 =$		25.	$8 \times 9 =$	
4.	$2 \times 9 =$		26.	$4 \times 9 =$	
5.	$9 \times 3 =$		27.	$7 \times 9 =$	
6.	$3 \times 9 =$		28.	$5 \times 9 =$	
7.	$9 \times 4 =$		29.	$6 \times 9 =$	
8.	$4 \times 9 =$		30.	$9 \times 5 =$	
9.	$9 \times 5 =$		31.	$9 \times 10 =$	
10.	$5 \times 9 =$		32.	$9 \times 1 =$	
11.	$9 \times 6 =$		33.	$9 \times 6 =$	
12.	$6 \times 9 =$		34.	$9 \times 4 =$	
13.	$9 \times 7 =$		35.	$9 \times 9 =$	
14.	$7 \times 9 =$		36.	$9 \times 2 =$	
15.	$9 \times 8 =$		37.	$9 \times 7 =$	
16.	$8 \times 9 =$		38.	$9 \times 3 =$	
17.	$9 \times 9 =$		39.	$9 \times 8 =$	
18.	$9 \times 10 =$		40.	$11 \times 9 =$	
19.	$10 \times 9 =$		41.	$9 \times 11 =$	
20.	$1 \times 9 =$		42.	$12 \times 9 =$	
21.	$10 \times 9 =$		43.	$9 \times 12 =$	
22.	$2 \times 9 =$		44.	$13 \times 9 =$	

B

Number Correct: _____

Multiply with Nine

Improvement: _____

1.	1 × 9 =		23.	10 × 9 =	
2.	9 × 1 =		24.	9 × 9 =	
3.	2 × 9 =		25.	4 × 9 =	
4.	9 × 2 =		26.	8 × 9 =	
5.	3 × 9 =		27.	3 × 9 =	
6.	9 × 3 =		28.	7 × 9 =	
7.	4 × 9 =		29.	6 × 9 =	
8.	9 × 4 =		30.	9 × 10 =	
9.	5 × 9 =		31.	9 × 5 =	
10.	9 × 5 =		32.	9 × 6 =	
11.	6 × 9 =		33.	9 × 1 =	
12.	9 × 6 =		34.	9 × 9 =	
13.	7 × 9 =		35.	9 × 4 =	
14.	9 × 7 =		36.	9 × 3 =	
15.	8 × 9 =		37.	9 × 2 =	
16.	9 × 8 =		38.	9 × 7 =	
17.	9 × 9 =		39.	9 × 8 =	
18.	10 × 9 =		40.	11 × 9 =	
19.	9 × 10 =		41.	9 × 11 =	
20.	9 × 3 =		42.	12 × 9 =	
21.	1 × 9 =		43.	9 × 12 =	
22.	2 × 9 =		44.	13 × 9 =	

Lesson 12:   Specify the corresponding whole when presented with one equal part.

33

© 2015 Great Minds®. eureka-math.org

A

Number Correct: _____

Multiply and Divide by Nine

1.	2 × 9 =	
2.	3 × 9 =	
3.	4 × 9 =	
4.	5 × 9 =	
5.	1 × 9 =	
6.	18 ÷ 9 =	
7.	27 ÷ 9 =	
8.	45 ÷ 9 =	
9.	9 ÷ 9 =	
10.	36 ÷ 9 =	
11.	6 × 9 =	
12.	7 × 9 =	
13.	8 × 9 =	
14.	9 × 9 =	
15.	10 × 9 =	
16.	72 ÷ 9 =	
17.	63 ÷ 9 =	
18.	81 ÷ 9 =	
19.	54 ÷ 9 =	
20.	90 ÷ 9 =	
21.	___ × 9 = 45	
22.	___ × 9 = 9	

23.	___ × 9 = 90	
24.	___ × 9 = 18	
25.	___ × 9 = 27	
26.	90 ÷ 9 =	
27.	45 ÷ 9 =	
28.	9 ÷ 9 =	
29.	18 ÷ 9 =	
30.	27 ÷ 9 =	
31.	___ × 9 = 54	
32.	___ × 9 = 63	
33.	___ × 9 = 81	
34.	___ × 9 = 72	
35.	63 ÷ 9 =	
36.	81 ÷ 9 =	
37.	54 ÷ 9 =	
38.	72 ÷ 9 =	
39.	11 × 9 =	
40.	99 ÷ 9 =	
41.	12 × 9 =	
42.	108 ÷ 9 =	
43.	14 × 9 =	
44.	126 ÷ 9 =	

Lesson 16:   Place whole number fractions and fractions between whole numbers on the number line.

35

© 2015 Great Minds®. eureka-math.org

B

Number Correct: _____

Multiply and Divide by Nine

Improvement: _____

1.	$1 \times 9 =$		23.	___ $\times 9 = 18$	
2.	$2 \times 9 =$		24.	___ $\times 9 = 90$	
3.	$3 \times 9 =$		25.	___ $\times 9 = 27$	
4.	$4 \times 9 =$		26.	$18 \div 9 =$	
5.	$5 \times 9 =$		27.	$9 \div 9 =$	
6.	$27 \div 9 =$		28.	$90 \div 9 =$	
7.	$18 \div 9 =$		29.	$45 \div 9 =$	
8.	$36 \div 9 =$		30.	$27 \div 9 =$	
9.	$9 \div 9 =$		31.	___ $\times 9 = 27$	
10.	$45 \div 9 =$		32.	___ $\times 9 = 36$	
11.	$10 \times 9 =$		33.	___ $\times 9 = 81$	
12.	$6 \times 9 =$		34.	___ $\times 9 = 63$	
13.	$7 \times 9 =$		35.	$72 \div 9 =$	
14.	$8 \times 9 =$		36.	$81 \div 9 =$	
15.	$9 \times 9 =$		37.	$54 \div 9 =$	
16.	$63 \div 9 =$		38.	$63 \div 9 =$	
17.	$54 \div 9 =$		39.	$11 \times 9 =$	
18.	$72 \div 9 =$		40.	$99 \div 9 =$	
19.	$90 \div 9 =$		41.	$12 \times 9 =$	
20.	$81 \div 9 =$		42.	$108 \div 9 =$	
21.	___ $\times 9 = 9$		43.	$13 \times 9 =$	
22.	___ $\times 9 = 45$		44.	$117 \div 9 =$	

Lesson 16: Place whole number fractions and fractions between whole numbers on the number line.

37

© 2015 Great Minds®. eureka-math.org

A

Number Correct: _____

Division

1.	3 ÷ 3 =		23.	24 ÷ 3 =	
2.	4 ÷ 4 =		24.	16 ÷ 2 =	
3.	5 ÷ 5 =		25.	30 ÷ 10 =	
4.	19 ÷ 19 =		26.	30 ÷ 3 =	
5.	0 ÷ 1 =		27.	27 ÷ 3 =	
6.	0 ÷ 2 =		28.	18 ÷ 2 =	
7.	0 ÷ 3 =		29.	40 ÷ 10 =	
8.	0 ÷ 19 =		30.	40 ÷ 4 =	
9.	6 ÷ 3 =		31.	20 ÷ 4 =	
10.	9 ÷ 3 =		32.	20 ÷ 5 =	
11.	12 ÷ 3 =		33.	24 ÷ 4 =	
12.	15 ÷ 3 =		34.	30 ÷ 5 =	
13.	4 ÷ 2 =		35.	28 ÷ 4 =	
14.	6 ÷ 2 =		36.	40 ÷ 5 =	
15.	8 ÷ 2 =		37.	32 ÷ 4 =	
16.	10 ÷ 2 =		38.	45 ÷ 5 =	
17.	18 ÷ 3 =		39.	44 ÷ 4 =	
18.	12 ÷ 2 =		40.	36 ÷ 4 =	
19.	21 ÷ 3 =		41.	48 ÷ 6 =	
20.	14 ÷ 2 =		42.	63 ÷ 7 =	
21.	20 ÷ 10 =		43.	64 ÷ 8 =	
22.	20 ÷ 2 =		44.	72 ÷ 9 =	

Lesson 17:   Practice placing various fractions on the number line.

39

© 2015 Great Minds®. eureka-math.org

B

Number Correct: _____

Improvement: _____

Division

1.	2 ÷ 2 =		23.	16 ÷ 2 =	
2.	3 ÷ 3 =		24.	24 ÷ 3 =	
3.	4 ÷ 4 =		25.	30 ÷ 3 =	
4.	17 ÷ 17 =		26.	30 ÷ 10 =	
5.	0 ÷ 2 =		27.	18 ÷ 2 =	
6.	0 ÷ 3 =		28.	27 ÷ 3 =	
7.	0 ÷ 4 =		29.	40 ÷ 4 =	
8.	0 ÷ 17 =		30.	40 ÷ 10 =	
9.	4 ÷ 2 =		31.	20 ÷ 5 =	
10.	6 ÷ 2 =		32.	20 ÷ 4 =	
11.	8 ÷ 2 =		33.	30 ÷ 5 =	
12.	10 ÷ 2 =		34.	24 ÷ 4 =	
13.	6 ÷ 3 =		35.	40 ÷ 5 =	
14.	9 ÷ 3 =		36.	28 ÷ 4 =	
15.	12 ÷ 3 =		37.	45 ÷ 5 =	
16.	15 ÷ 3 =		38.	32 ÷ 4 =	
17.	12 ÷ 2 =		39.	55 ÷ 5 =	
18.	18 ÷ 3 =		40.	36 ÷ 4 =	
19.	14 ÷ 2 =		41.	54 ÷ 6 =	
20.	21 ÷ 3 =		42.	56 ÷ 7 =	
21.	20 ÷ 2 =		43.	72 ÷ 8 =	
22.	20 ÷ 10 =		44.	63 ÷ 9 =	

Lesson 17:   Practice placing various fractions on the number line.

41

© 2013 Great Minds . eureka-math.org

A

Number Correct: _____

Express Fractions as Whole Numbers

1.	$2/1 =$		23.	$6/3 =$	
2.	$2/2 =$		24.	$3/3 =$	
3.	$4/2 =$		25.	$3/1 =$	
4.	$6/2 =$		26.	$9/3 =$	
5.	$10/2 =$		27.	$16/4 =$	
6.	$8/2 =$		28.	$20/4 =$	
7.	$5/1 =$		29.	$12/3 =$	
8.	$5/5 =$		30.	$15/3 =$	
9.	$10/5 =$		31.	$70/10 =$	
10.	$15/5 =$		32.	$12/2 =$	
11.	$25/5 =$		33.	$14/2 =$	
12.	$20/5 =$		34.	$90/10 =$	
13.	$10/10 =$		35.	$30/5 =$	
14.	$50/10 =$		36.	$35/5 =$	
15.	$30/10 =$		37.	$60/10 =$	
16.	$10/1 =$		38.	$18/2 =$	
17.	$20/10 =$		39.	$40/5 =$	
18.	$40/10 =$		40.	$80/10 =$	
19.	$8/4 =$		41.	$16/2 =$	
20.	$4/4 =$		42.	$45/5 =$	
21.	$4/1 =$		43.	$27/3 =$	
22.	$12/4 =$		44.	$32/4 =$	

Lesson 19:   Understand distance and position on the number line as strategies for comparing fractions. (Optional)

43

© 2015 Great Minds® eureka-math.org

B

Number Correct: _____

Improvement: _____

Express Fractions as Whole Numbers

1.	$5/1 =$		23.	$8/4 =$	
2.	$5/5 =$		24.	$4/4 =$	
3.	$10/5 =$		25.	$4/1 =$	
4.	$15/5 =$		26.	$12/4 =$	
5.	$25/5 =$		27.	$12/3 =$	
6.	$20/5 =$		28.	$15/3 =$	
7.	$2/1 =$		29.	$16/4 =$	
8.	$2/2 =$		30.	$20/4 =$	
9.	$4/2 =$		31.	$90/10 =$	
10.	$6/2 =$		32.	$30/5 =$	
11.	$10/2 =$		33.	$35/5 =$	
12.	$8/2 =$		34.	$70/10 =$	
13.	$10/1 =$		35.	$12/2 =$	
14.	$10/10 =$		36.	$14/2 =$	
15.	$50/10 =$		37.	$80/10 =$	
16.	$30/10 =$		38.	$45/5 =$	
17.	$20/10 =$		39.	$16/2 =$	
18.	$40/10 =$		40.	$60/10 =$	
19.	$6/3 =$		41.	$18/2 =$	
20.	$3/3 =$		42.	$40/5 =$	
21.	$3/1 =$		43.	$36/4 =$	
22.	$9/3 =$		44.	$24/3 =$	

Lesson 19: Understand distance and position on the number line as strategies for comparing fractions. (Optional)

© 2015 Great Minds®. eureka-math.org

45

Multiply.

7 × 1 = _____   7 × 2 = _____   7 × 3 = _____   7 × 4 = _____

7 × 5 = _____   7 × 1 = _____   7 × 2 = _____   7 × 1 = _____

7 × 3 = _____   7 × 1 = _____   7 × 4 = _____   7 × 1 = _____

7 × 5 = _____   7 × 1 = _____   7 × 2 = _____   7 × 3 = _____

7 × 2 = _____   7 × 4 = _____   7 × 2 = _____   7 × 5 = _____

7 × 2 = _____   7 × 1 = _____   7 × 2 = _____   7 × 3 = _____

7 × 1 = _____   7 × 3 = _____   7 × 2 = _____   7 × 3 = _____

7 × 4 = _____   7 × 3 = _____   7 × 5 = _____   7 × 3 = _____

7 × 4 = _____   7 × 1 = _____   7 × 4 = _____   7 × 2 = _____

7 × 4 = _____   7 × 3 = _____   7 × 4 = _____   7 × 5 = _____

7 × 4 = _____   7 × 5 = _____   7 × 1 = _____   7 × 5 = _____

7 × 2 = _____   7 × 5 = _____   7 × 3 = _____   7 × 5 = _____

7 × 4 = _____   7 × 2 = _____   7 × 4 = _____   7 × 3 = _____

7 × 5 = _____   7 × 3 = _____   7 × 2 = _____   7 × 4 = _____

7 × 3 = _____   7 × 5 = _____   7 × 2 = _____   7 × 4 = _____

_____

multiply by 7 (1–5)

Lesson 20:   Recognize and show that equivalent fractions have the same size, though not necessarily the same shape.

47

© 2015 Great Minds®. eureka-math.org

A

Number Correct: _____

Add by Six

1.	0 + 6 =	
2.	1 + 6 =	
3.	2 + 6 =	
4.	3 + 6 =	
5.	4 + 6 =	
6.	6 + 4 =	
7.	6 + 3 =	
8.	6 + 2 =	
9.	6 + 1 =	
10.	6 + 0 =	
11.	15 + 6 =	
12.	25 + 6 =	
13.	35 + 6 =	
14.	45 + 6 =	
15.	55 + 6 =	
16.	85 + 6 =	
17.	6 + 6 =	
18.	16 + 6 =	
19.	26 + 6 =	
20.	36 + 6 =	
21.	46 + 6 =	
22.	76 + 6 =	

23.	7 + 6 =	
24.	17 + 6 =	
25.	27 + 6 =	
26.	37 + 6 =	
27.	47 + 6 =	
28.	77 + 6 =	
29.	8 + 6 =	
30.	18 + 6 =	
31.	28 + 6 =	
32.	38 + 6 =	
33.	48 + 6 =	
34.	78 + 6 =	
35.	9 + 6 =	
36.	19 + 6 =	
37.	29 + 6 =	
38.	39 + 6 =	
39.	89 + 6 =	
40.	6 + 75 =	
41.	6 + 56 =	
42.	6 + 77 =	
43.	6 + 88 =	
44.	6 + 99 =	

Lesson 23: Generate simple equivalent fractions by using visual fraction models and the number line.

© 2015 Great Minds®. eureka-math.org

B

Number Correct: _____

Add by Six

Improvement: _____

1.	6 + 0 =	
2.	6 + 1 =	
3.	6 + 2 =	
4.	6 + 3 =	
5.	6 + 4 =	
6.	4 + 6 =	
7.	3 + 6 =	
8.	2 + 6 =	
9.	1 + 6 =	
10.	0 + 6 =	
11.	5 + 6 =	
12.	15 + 6 =	
13.	25 + 6 =	
14.	35 + 6 =	
15.	45 + 6 =	
16.	75 + 6 =	
17.	6 + 6 =	
18.	16 + 6 =	
19.	26 + 6 =	
20.	36 + 6 =	
21.	46 + 6 =	
22.	86 + 6 =	

23.	7 + 6 =	
24.	17 + 6 =	
25.	27 + 6 =	
26.	37 + 6 =	
27.	47 + 6 =	
28.	67 + 6 =	
29.	8 + 6 =	
30.	18 + 6 =	
31.	28 + 6 =	
32.	38 + 6 =	
33.	48 + 6 =	
34.	88 + 6 =	
35.	9 + 6 =	
36.	19 + 6 =	
37.	29 + 6 =	
38.	39 + 6 =	
39.	79 + 6 =	
40.	6 + 55 =	
41.	6 + 76 =	
42.	6 + 57 =	
43.	6 + 98 =	
44.	6 + 89 =	

Lesson 23: Generate simple equivalent fractions by using visual fraction models and the number line.

51

© 2015 Great Minds®. eureka-math.org

A

Number Correct: _____

Add by Seven

1.	0 + 7 =	
2.	1 + 7 =	
3.	2 + 7 =	
4.	3 + 7 =	
5.	7 + 3 =	
6.	7 + 2 =	
7.	7 + 1 =	
8.	7 + 0 =	
9.	4 + 7 =	
10.	14 + 7 =	
11.	24 + 7 =	
12.	34 + 7 =	
13.	44 + 7 =	
14.	84 + 7 =	
15.	64 + 7 =	
16.	5 + 7 =	
17.	15 + 7 =	
18.	25 + 7 =	
19.	35 + 7 =	
20.	45 + 7 =	
21.	75 + 7 =	
22.	55 + 7 =	

23.	6 + 7 =	
24.	16 + 7 =	
25.	26 + 7 =	
26.	36 + 7 =	
27.	46 + 7 =	
28.	66 + 7 =	
29.	7 + 7 =	
30.	17 + 7 =	
31.	27 + 7 =	
32.	37 + 7 =	
33.	87 + 7 =	
34.	8 + 7 =	
35.	18 + 7 =	
36.	28 + 7 =	
37.	38 + 7 =	
38.	78 + 7 =	
39.	9 + 7 =	
40.	19 + 7 =	
41.	29 + 7 =	
42.	39 + 7 =	
43.	49 + 7 =	
44.	79 + 7 =	

Lesson 24:   Express whole numbers as fractions and recognize equivalence with different units.

53

© 2015 Great Minds®. eureka-math.org

B

Add by Seven

Number Correct: _____

Improvement: _____

1.	7 + 0 =	
2.	7 + 1 =	
3.	7 + 2 =	
4.	7 + 3 =	
5.	3 + 7 =	
6.	2 + 7 =	
7.	1 + 7 =	
8.	0 + 7 =	
9.	4 + 7 =	
10.	14 + 7 =	
11.	24 + 7 =	
12.	34 + 7 =	
13.	44 + 7 =	
14.	74 + 7 =	
15.	54 + 7 =	
16.	5 + 7 =	
17.	15 + 7 =	
18.	25 + 7 =	
19.	35 + 7 =	
20.	45 + 7 =	
21.	85 + 7 =	
22.	65 + 7 =	

23.	6 + 7 =	
24.	16 + 7 =	
25.	26 + 7 =	
26.	36 + 7 =	
27.	46 + 7 =	
28.	76 + 7 =	
29.	7 + 7 =	
30.	17 + 7 =	
31.	27 + 7 =	
32.	37 + 7 =	
33.	67 + 7 =	
34.	8 + 7 =	
35.	18 + 7 =	
36.	28 + 7 =	
37.	38 + 7 =	
38.	88 + 7 =	
39.	9 + 7 =	
40.	19 + 7 =	
41.	29 + 7 =	
42.	39 + 7 =	
43.	49 + 7 =	
44.	89 + 7 =	

Lesson 24:   Express whole numbers as fractions and recognize equivalence with different units.

© 2015 Great Minds®. eureka-math.org

55

A

Number Correct: _____

Subtract by Six

1.	16 – 6 =	
2.	6 – 6 =	
3.	26 – 6 =	
4.	7 – 6 =	
5.	17 – 6 =	
6.	37 – 6 =	
7.	8 – 6 =	
8.	18 – 6 =	
9.	48 – 6 =	
10.	9 – 6 =	
11.	19 – 6 =	
12.	59 – 6 =	
13.	10 – 6 =	
14.	20 – 6 =	
15.	70 – 6 =	
16.	11 – 6 =	
17.	21 – 6 =	
18.	81 – 6 =	
19.	12 – 6 =	
20.	22 – 6 =	
21.	82 – 6 =	
22.	13 – 6 =	

23.	23 – 6 =	
24.	33 – 6 =	
25.	63 – 6 =	
26.	83 – 6 =	
27.	14 – 6 =	
28.	24 – 6 =	
29.	34 – 6 =	
30.	74 – 6 =	
31.	54 – 6 =	
32.	15 – 6 =	
33.	25 – 6 =	
34.	35 – 6 =	
35.	85 – 6 =	
36.	65 – 6 =	
37.	90 – 6 =	
38.	53 – 6 =	
39.	42 – 6 =	
40.	71 – 6 =	
41.	74 – 6 =	
42.	95 – 6 =	
43.	51 – 6 =	
44.	92 – 6 =	

© 2015 Great Minds®. eureka-math.org

B

Number Correct: _____

Subtract by Six

Improvement: _____

1.	6 − 6 =	
2.	16 − 6 =	
3.	26 − 6 =	
4.	7 − 6 =	
5.	17 − 6 =	
6.	67 − 6 =	
7.	8 − 6 =	
8.	18 − 6 =	
9.	78 − 6 =	
10.	9 − 6 =	
11.	19 − 6 =	
12.	89 − 6 =	
13.	10 − 6 =	
14.	20 − 6 =	
15.	90 − 6 =	
16.	11 − 6 =	
17.	21 − 6 =	
18.	41 − 6 =	
19.	12 − 6 =	
20.	22 − 6 =	
21.	42 − 6 =	
22.	13 − 6 =	

23.	23 − 6 =	
24.	33 − 6 =	
25.	53 − 6 =	
26.	73 − 6 =	
27.	14 − 6 =	
28.	24 − 6 =	
29.	34 − 6 =	
30.	64 − 6 =	
31.	44 − 6 =	
32.	15 − 6 =	
33.	25 − 6 =	
34.	35 − 6 =	
35.	75 − 6 =	
36.	55 − 6 =	
37.	70 − 6 =	
38.	63 − 6 =	
39.	52 − 6 =	
40.	81 − 6 =	
41.	64 − 6 =	
42.	85 − 6 =	
43.	91 − 6 =	
44.	52 − 6 =	

EUREKA MATH

Lesson 25: Express whole number fractions on the number line when the unit interval is 1.

59

© 2015 Great Minds®. eureka-math.org

A

Number Correct: _____

Add by Eight

1.	0 + 8 =	
2.	1 + 8 =	
3.	2 + 8 =	
4.	8 + 2 =	
5.	1 + 8 =	
6.	0 + 8 =	
7.	3 + 8 =	
8.	13 + 8 =	
9.	23 + 8 =	
10.	33 + 8 =	
11.	43 + 8 =	
12.	83 + 8 =	
13.	4 + 8 =	
14.	14 + 8 =	
15.	24 + 8 =	
16.	34 + 8 =	
17.	44 + 8 =	
18.	74 + 8 =	
19.	5 + 8 =	
20.	15 + 8 =	
21.	25 + 8 =	
22.	35 + 8 =	

23.	65 + 8 =	
24.	6 + 8 =	
25.	16 + 8 =	
26.	26 + 8 =	
27.	36 + 8 =	
28.	86 + 8 =	
29.	46 + 8 =	
30.	7 + 8 =	
31.	17 + 8 =	
32.	27 + 8 =	
33.	37 + 8 =	
34.	77 + 8 =	
35.	8 + 8 =	
36.	18 + 8 =	
37.	28 + 8 =	
38.	38 + 8 =	
39.	68 + 8 =	
40.	9 + 8 =	
41.	19 + 8 =	
42.	29 + 8 =	
43.	39 + 8 =	
44.	89 + 8 =	

EUREKA MATH®

Lesson 26: Decompose whole number fractions greater than 1 using whole number equivalence with various models.

61

© 2015 Great Minds®. eureka-math.org

B

Number Correct: _____

Add by Eight

Improvement: _____

1.	8 + 0 =		23.	55 + 8 =	
2.	8 + 1 =		24.	6 + 8 =	
3.	8 + 2 =		25.	16 + 8 =	
4.	2 + 8 =		26.	26 + 8 =	
5.	1 + 8 =		27.	36 + 8 =	
6.	0 + 8 =		28.	66 + 8 =	
7.	3 + 8 =		29.	56 + 8 =	
8.	13 + 8 =		30.	7 + 8 =	
9.	23 + 8 =		31.	17 + 8 =	
10.	33 + 8 =		32.	27 + 8 =	
11.	43 + 8 =		33.	37 + 8 =	
12.	73 + 8 =		34.	67 + 8 =	
13.	4 + 8 =		35.	8 + 8 =	
14.	14 + 8 =		36.	18 + 8 =	
15.	24 + 8 =		37.	28 + 8 =	
16.	34 + 8 =		38.	38 + 8 =	
17.	44 + 8 =		39.	78 + 8 =	
18.	84 + 8 =		40.	9 + 8 =	
19.	5 + 8 =		41.	19 + 8 =	
20.	15 + 8 =		42.	29 + 8 =	
21.	25 + 8 =		43.	39 + 8 =	
22.	35 + 8 =		44.	89 + 8 =	

Lesson 26:   Decompose whole number fractions greater than 1 using whole number equivalence with various models.

63

© 2015 Great Minds®. eureka-math.org

A

Number Correct: _____

Subtract by Seven

1.	17 – 7 =	
2.	7 – 7 =	
3.	27 – 7 =	
4.	8 – 7 =	
5.	18 – 7 =	
6.	38 – 7 =	
7.	9 – 7 =	
8.	19 – 7 =	
9.	49 – 7 =	
10.	10 – 7 =	
11.	20 – 7 =	
12.	60 – 7 =	
13.	11 – 7 =	
14.	21 – 7 =	
15.	71 – 7 =	
16.	12 – 7 =	
17.	22 – 7 =	
18.	82 – 7 =	
19.	13 – 7 =	
20.	23 – 7 =	
21.	83 – 7 =	
22.	14 – 7 =	

23.	24 – 7 =	
24.	34 – 7 =	
25.	64 – 7 =	
26.	84 – 7 =	
27.	15 – 7 =	
28.	25 – 7 =	
29.	35 – 7 =	
30.	75 – 7 =	
31.	55 – 7 =	
32.	16 – 7 =	
33.	26 – 7 =	
34.	36 – 7 =	
35.	86 – 7 =	
36.	66 – 7 =	
37.	90 – 7 =	
38.	53 – 7 =	
39.	42 – 7 =	
40.	71 – 7 =	
41.	74 – 7 =	
42.	56 – 7 =	
43.	95 – 7 =	
44.	92 – 7 =	

Lesson 27: Explain equivalence by manipulating units and reasoning about their size.

65

© 2015 Great Minds®. eureka-math.org

B

Number Correct: _____

Subtract by Seven

Improvement: _____

1.	7 – 7 =	
2.	17 – 7 =	
3.	27 – 7 =	
4.	8 – 7 =	
5.	18 – 7 =	
6.	68 – 7 =	
7.	9 – 7 =	
8.	19 – 7 =	
9.	79 – 7 =	
10.	10 – 7 =	
11.	20 – 7 =	
12.	90 – 7 =	
13.	11 – 7 =	
14.	21 – 7 =	
15.	91 – 7 =	
16.	12 – 7 =	
17.	22 – 7 =	
18.	42 – 7 =	
19.	13 – 7 =	
20.	23 – 7 =	
21.	43 – 7 =	
22.	14 – 7 =	

23.	24 – 7 =	
24.	34 – 7 =	
25.	54 – 7 =	
26.	74 – 7 =	
27.	15 – 7 =	
28.	25 – 7 =	
29.	35 – 7 =	
30.	65 – 7 =	
31.	45 – 7 =	
32.	16 – 7 =	
33.	26 – 7 =	
34.	36 – 7 =	
35.	76 – 7 =	
36.	56 – 7 =	
37.	70 – 7 =	
38.	63 – 7 =	
39.	52 – 7 =	
40.	81 – 7 =	
41.	74 – 7 =	
42.	66 – 7 =	
43.	85 – 7 =	
44.	52 – 7 =	

Lesson 27: Explain equivalence by manipulating units and reasoning about their size.

67

© 2015 Great Minds®. eureka-math.org

A

Number Correct: _____

Subtract by Eight

1.	18 – 8 =	
2.	8 – 8 =	
3.	28 – 8 =	
4.	9 – 8 =	
5.	19 – 8 =	
6.	39 – 8 =	
7.	10 – 8 =	
8.	20 – 8 =	
9.	50 – 8 =	
10.	11 – 8 =	
11.	21 – 8 =	
12.	71 – 8 =	
13.	12 – 8 =	
14.	22 – 8 =	
15.	82 – 8 =	
16.	13 – 8 =	
17.	23 – 8 =	
18.	83 – 8 =	
19.	14 – 8 =	
20.	24 – 8 =	
21.	34 – 8 =	
22.	54 – 8 =	

23.	74 – 8 =	
24.	15 – 8 =	
25.	25 – 8 =	
26.	35 – 8 =	
27.	85 – 8 =	
28.	65 – 8 =	
29.	16 – 8 =	
30.	26 – 8 =	
31.	36 – 8 =	
32.	96 – 8 =	
33.	76 – 8 =	
34.	17 – 8 =	
35.	27 – 8 =	
36.	37 – 8 =	
37.	87 – 8 =	
38.	67 – 8 =	
39.	70 – 8 =	
40.	62 – 8 =	
41.	84 – 8 =	
42.	66 – 8 =	
43.	91 – 8 =	
44.	75 – 8 =	

Lesson 28: Compare fractions with the same numerator pictorially.

© 2015 Great Minds®. eureka-math.org

B

Number Correct: _____

Improvement: _____

Subtract by Eight

1.	8 – 8 =	
2.	18 – 8 =	
3.	28 – 8 =	
4.	9 – 8 =	
5.	19 – 8 =	
6.	69 – 8 =	
7.	10 – 8 =	
8.	20 – 8 =	
9.	60 – 8 =	
10.	11 – 8 =	
11.	21 – 8 =	
12.	81 – 8 =	
13.	12 – 8 =	
14.	22 – 8 =	
15.	52 – 8 =	
16.	13 – 8 =	
17.	23 – 8 =	
18.	93 – 8 =	
19.	14 – 8 =	
20.	24 – 8 =	
21.	34 – 8 =	
22.	74 – 8 =	

23.	94 – 8 =	
24.	15 – 8 =	
25.	25 – 8 =	
26.	35 – 8 =	
27.	95 – 8 =	
28.	75 – 8 =	
29.	16 – 8 =	
30.	26 – 8 =	
31.	36 – 8 =	
32.	66 – 8 =	
33.	46 – 8 =	
34.	17 – 8 =	
35.	27 – 8 =	
36.	37 – 8 =	
37.	97 – 8 =	
38.	77 – 8 =	
39.	80 – 8 =	
40.	71 – 8 =	
41.	53 – 8 =	
42.	45 – 8 =	
43.	87 – 8 =	
44.	54 – 8 =	

© 2015 Great Minds®. eureka-math.org

Multiply.

8 × 1 = _____	8 × 2 = _____	8 × 3 = _____	8 × 4 = _____
8 × 5 = _____	8 × 6 = _____	8 × 7 = _____	8 × 8 = _____
8 × 9 = _____	8 × 10 = _____	8 × 5 = _____	8 × 6 = _____
8 × 5 = _____	8 × 7 = _____	8 × 5 = _____	8 × 8 = _____
8 × 5 = _____	8 × 9 = _____	8 × 5 = _____	8 × 10 = _____
8 × 6 = _____	8 × 5 = _____	8 × 6 = _____	8 × 7 = _____
8 × 6 = _____	8 × 8 = _____	8 × 6 = _____	8 × 9 = _____
8 × 6 = _____	8 × 7 = _____	8 × 6 = _____	8 × 7 = _____
8 × 8 = _____	8 × 7 = _____	8 × 9 = _____	8 × 7 = _____
8 × 8 = _____	8 × 6 = _____	8 × 8 = _____	8 × 7 = _____
8 × 8 = _____	8 × 9 = _____	8 × 9 = _____	8 × 6 = _____
8 × 9 = _____	8 × 7 = _____	8 × 9 = _____	8 × 8 = _____
8 × 9 = _____	8 × 8 = _____	8 × 6 = _____	8 × 9 = _____
8 × 7 = _____	8 × 9 = _____	8 × 6 = _____	8 × 8 = _____
8 × 9 = _____	8 × 7 = _____	8 × 6 = _____	8 × 8 = _____

multiply by 8 (5–9)

Lesson 29:   Compare fractions with the same numerator using <, >, or =, and use a
model to reason about their size.

73

© 2015 Great Minds®. eureka-math.org

Multiply.

9 × 1 = _____   9 × 2 = _____   9 × 3 = _____   9 × 4 = _____

9 × 5 = _____   9 × 1 = _____   9 × 2 = _____   9 × 1 = _____

9 × 3 = _____   9 × 1 = _____   9 × 4 = _____   9 × 1 = _____

9 × 5 = _____   9 × 1 = _____   9 × 2 = _____   9 × 3 = _____

9 × 2 = _____   9 × 4 = _____   9 × 2 = _____   9 × 5 = _____

9 × 2 = _____   9 × 1 = _____   9 × 2 = _____   9 × 3 = _____

9 × 1 = _____   9 × 3 = _____   9 × 2 = _____   9 × 3 = _____

9 × 4 = _____   9 × 3 = _____   9 × 5 = _____   9 × 3 = _____

9 × 4 = _____   9 × 1 = _____   9 × 4 = _____   9 × 2 = _____

9 × 4 = _____   9 × 3 = _____   9 × 4 = _____   9 × 5 = _____

9 × 4 = _____   9 × 5 = _____   9 × 1 = _____   9 × 5 = _____

9 × 2 = _____   9 × 5 = _____   9 × 3 = _____   9 × 5 = _____

9 × 4 = _____   9 × 2 = _____   9 × 4 = _____   9 × 3 = _____

9 × 5 = _____   9 × 3 = _____   9 × 2 = _____   9 × 4 = _____

9 × 3 = _____   9 × 5 = _____   9 × 2 = _____   9 × 4 = _____

multiply by 9 (1–5)

Lesson 30:   Partition various wholes precisely into equal parts using a number method.

75

© 2015 Great Minds®. eureka-math.org

Grade 3
Module 6

A

Number Correct: _____

Multiply or Divide by 6

1.	2 × 6 =	
2.	3 × 6 =	
3.	4 × 6 =	
4.	5 × 6 =	
5.	1 × 6 =	
6.	12 ÷ 6 =	
7.	18 ÷ 6 =	
8.	30 ÷ 6 =	
9.	6 ÷ 6 =	
10.	24 ÷ 6 =	
11.	6 × 6 =	
12.	7 × 6 =	
13.	8 × 6 =	
14.	9 × 6 =	
15.	10 × 6 =	
16.	48 ÷ 6 =	
17.	42 ÷ 6 =	
18.	54 ÷ 6 =	
19.	36 ÷ 6 =	
20.	60 ÷ 6 =	
21.	____ × 6 = 30	
22.	____ × 6 = 6	

23.	____ × 6 = 60	
24.	____ × 6 = 12	
25.	____ × 6 = 18	
26.	60 ÷ 6 =	
27.	30 ÷ 6 =	
28.	6 ÷ 6 =	
29.	12 ÷ 6 =	
30.	18 ÷ 6 =	
31.	____ × 6 = 36	
32.	____ × 6 = 42	
33.	____ × 6 = 54	
34.	____ × 6 = 48	
35.	42 ÷ 6 =	
36.	54 ÷ 6 =	
37.	36 ÷ 6 =	
38.	48 ÷ 6 =	
39.	11 × 6 =	
40.	66 ÷ 6 =	
41.	12 × 6 =	
42.	72 ÷ 6 =	
43.	14 × 6 =	
44.	84 ÷ 6 =	

Lesson 3:   Create scaled bar graphs.

79

© 2015 Great Minds®. eureka-math.org

B

Number Correct: _____

Multiply or Divide by 6

Improvement: _____

1.	1 × 6 =	
2.	2 × 6 =	
3.	3 × 6 =	
4.	4 × 6 =	
5.	5 × 6 =	
6.	18 ÷ 6 =	
7.	12 ÷ 6 =	
8.	24 ÷ 6 =	
9.	6 ÷ 6 =	
10.	30 ÷ 6 =	
11.	10 × 6 =	
12.	6 × 6 =	
13.	7 × 6 =	
14.	8 × 6 =	
15.	9 × 6 =	
16.	42 ÷ 6 =	
17.	36 ÷ 6 =	
18.	48 ÷ 6 =	
19.	60 ÷ 6 =	
20.	54 ÷ 6 =	
21.	____ × 6 = 6	
22.	____ × 6 = 30	

23.	____ × 6 = 12	
24.	____ × 6 = 60	
25.	____ × 6 = 18	
26.	12 ÷ 6 =	
27.	6 ÷ 6 =	
28.	60 ÷ 6 =	
29.	30 ÷ 6 =	
30.	18 ÷ 6 =	
31.	____ × 6 = 18	
32.	____ × 6 = 24	
33.	____ × 6 = 54	
34.	____ × 6 = 42	
35.	48 ÷ 6 =	
36.	54 ÷ 6 =	
37.	36 ÷ 6 =	
38.	42 ÷ 6 =	
39.	11 × 6 =	
40.	66 ÷ 6 =	
41.	12 × 6 =	
42.	72 ÷ 6 =	
43.	13 × 6 =	
44.	78 ÷ 6 =	

Lesson 3:   Create scaled bar graphs.

81

© 2015 Great Minds®. eureka-math.org

Number of Children in Third-Grade Families

```
                    X
  X                 X
  X                 X
  X                 X         X
  X                 X         X
  X                 X         X
  X                 X         X
  X                 X         X         X
  X                 X         X         X
─────────────────────────────────────────────
  1                 2         3         4
```

Number of Children   **X** = *1 Child*

_____

line plot

Lesson 4:   Solve one- and two-step problems involving graphs.

83

© 2015 Great Minds®. eureka-math.org

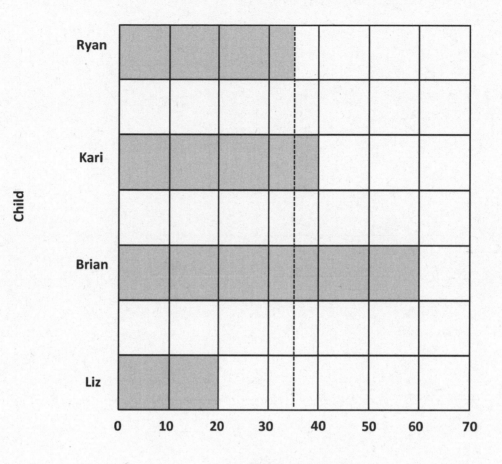

Number of Minutes Spent Practicing Piano

bar graph

Lesson 4: Solve one- and two-step problems involving graphs.

85

© 2015 Great Minds®. eureka-math.org

Multiply.

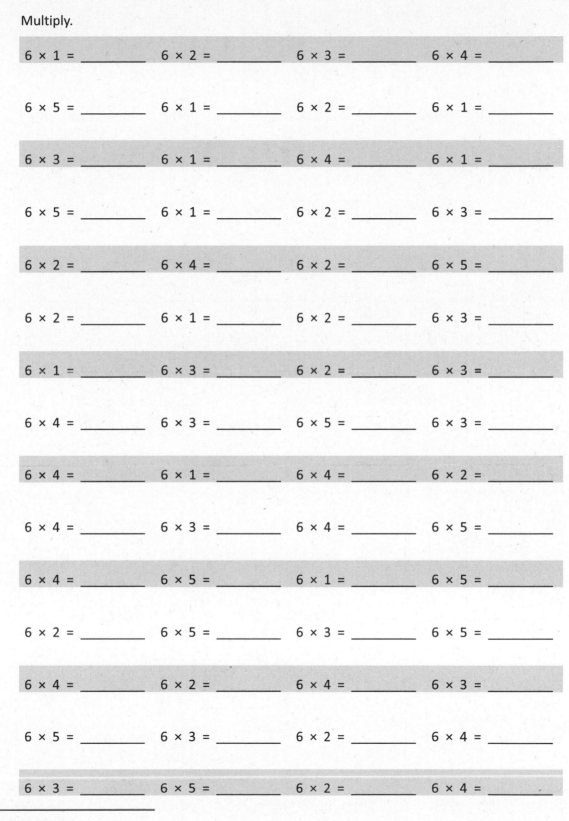

6 × 1 = _____   6 × 2 = _____   6 × 3 = _____   6 × 4 = _____

6 × 5 = _____   6 × 1 = _____   6 × 2 = _____   6 × 1 = _____

6 × 3 = _____   6 × 1 = _____   6 × 4 = _____   6 × 1 = _____

6 × 5 = _____   6 × 1 = _____   6 × 2 = _____   6 × 3 = _____

6 × 2 = _____   6 × 4 = _____   6 × 2 = _____   6 × 5 = _____

6 × 2 = _____   6 × 1 = _____   6 × 2 = _____   6 × 3 = _____

6 × 1 = _____   6 × 3 = _____   6 × 2 = _____   6 × 3 = _____

6 × 4 = _____   6 × 3 = _____   6 × 5 = _____   6 × 3 = _____

6 × 4 = _____   6 × 1 = _____   6 × 4 = _____   6 × 2 = _____

6 × 4 = _____   6 × 3 = _____   6 × 4 = _____   6 × 5 = _____

6 × 4 = _____   6 × 5 = _____   6 × 1 = _____   6 × 5 = _____

6 × 2 = _____   6 × 5 = _____   6 × 3 = _____   6 × 5 = _____

6 × 4 = _____   6 × 2 = _____   6 × 4 = _____   6 × 3 = _____

6 × 5 = _____   6 × 3 = _____   6 × 2 = _____   6 × 4 = _____

6 × 3 = _____   6 × 5 = _____   6 × 2 = _____   6 × 4 = _____

multiply by 6 (1–5)

Number of Miles a Truck Driver Drives

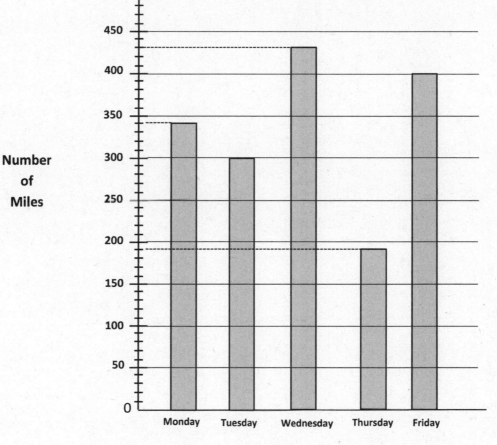

Number of Miles (y-axis): 0, 50, 100, 150, 200, 250, 300, 350, 400, 450, 500

Day (x-axis): Monday, Tuesday, Wednesday, Thursday, Friday

number of miles bar graph

Multiply.

6 × 1 = _____	6 × 2 = _____	6 × 3 = _____	6 × 4 = _____
6 × 5 = _____	6 × 6 = _____	6 × 7 = _____	6 × 8 = _____
6 × 9 = _____	6 × 10 = _____	6 × 5 = _____	6 × 6 = _____
6 × 5 = _____	6 × 7 = _____	6 × 5 = _____	6 × 8 = _____
6 × 5 = _____	6 × 9 = _____	6 × 5 = _____	6 × 10 = _____
6 × 6 = _____	6 × 5 = _____	6 × 6 = _____	6 × 7 = _____
6 × 6 = _____	6 × 8 = _____	6 × 6 = _____	6 × 9 = _____
6 × 6 = _____	6 × 7 = _____	6 × 6 = _____	6 × 7 = _____
6 × 8 = _____	6 × 7 = _____	6 × 9 = _____	6 × 7 = _____
6 × 8 = _____	6 × 6 = _____	6 × 8 = _____	6 × 7 = _____
6 × 8 = _____	6 × 9 = _____	6 × 9 = _____	6 × 6 = _____
6 × 9 = _____	6 × 7 = _____	6 × 9 = _____	6 × 8 = _____
6 × 9 = _____	6 × 8 = _____	6 × 6 = _____	6 × 9 = _____
6 × 7 = _____	6 × 9 = _____	6 × 6 = _____	6 × 8 = _____
6 × 9 = _____	6 × 7 = _____	6 × 6 = _____	6 × 8 = _____

multiply by 6 (6–10)

© 2015 Great Minds®. eureka-math.org

Multiply.

7 × 1 = _____   7 × 2 = _____   7 × 3 = _____   7 × 4 = _____

7 × 5 = _____   7 × 1 = _____   7 × 2 = _____   7 × 1 = _____

7 × 3 = _____   7 × 1 = _____   7 × 4 = _____   7 × 1 = _____

7 × 5 = _____   7 × 1 = _____   7 × 2 = _____   7 × 3 = _____

7 × 2 = _____   7 × 4 = _____   7 × 2 = _____   7 × 5 = _____

7 × 2 = _____   7 × 1 = _____   7 × 2 = _____   7 × 3 = _____

7 × 1 = _____   7 × 3 = _____   7 × 2 = _____   7 × 3 = _____

7 × 4 = _____   7 × 3 = _____   7 × 5 = _____   7 × 3 = _____

7 × 4 = _____   7 × 1 = _____   7 × 4 = _____   7 × 2 = _____

7 × 4 = _____   7 × 3 = _____   7 × 4 = _____   7 × 5 = _____

7 × 4 = _____   7 × 5 = _____   7 × 1 = _____   7 × 5 = _____

7 × 2 = _____   7 × 5 = _____   7 × 3 = _____   7 × 5 = _____

7 × 4 = _____   7 × 2 = _____   7 × 4 = _____   7 × 3 = _____

7 × 5 = _____   7 × 3 = _____   7 × 2 = _____   7 × 4 = _____

7 × 3 = _____   7 × 5 = _____   7 × 2 = _____   7 × 4 = _____

multiply by 7 (1–5)

EUREKA MATH®

© 2015 Great Minds®. eureka-math.org

Multiply.

7 × 1 = _____   7 × 2 = _____   7 × 3 = _____   7 × 4 = _____

7 × 5 = _____   7 × 6 = _____   7 × 7 = _____   7 × 8 = _____

7 × 9 = _____   7 × 10 = _____   7 × 5 = _____   7 × 6 = _____

7 × 5 = _____   7 × 7 = _____   7 × 5 = _____   7 × 8 = _____

7 × 5 = _____   7 × 9 = _____   7 × 5 = _____   7 × 10 = _____

7 × 6 = _____   7 × 5 = _____   7 × 6 = _____   7 × 7 = _____

7 × 6 = _____   7 × 8 = _____   7 × 6 = _____   7 × 9 = _____

7 × 6 = _____   7 × 7 = _____   7 × 6 = _____   7 × 7 = _____

7 × 8 = _____   7 × 7 = _____   7 × 9 = _____   7 × 7 = _____

7 × 8 = _____   7 × 6 = _____   7 × 8 = _____   7 × 7 = _____

7 × 8 = _____   7 × 9 = _____   7 × 9 = _____   7 × 6 = _____

7 × 9 = _____   7 × 7 = _____   7 × 9 = _____   7 × 8 = _____

7 × 9 = _____   7 × 8 = _____   7 × 6 = _____   7 × 9 = _____

7 × 7 = _____   7 × 9 = _____   7 × 6 = _____   7 × 8 = _____

7 × 9 = _____   7 × 7 = _____   7 × 6 = _____   7 × 8 = _____

multiply by 7 (6–10)

Lesson 9:   Analyze data to problem solve.   95

© 2015 Great Minds®. eureka-math.org

Grade 3
Module 7

Multiply.

3 × 1 = _____   3 × 2 = _____   3 × 3 = _____   3 × 4 = _____

3 × 5 = _____   3 × 1 = _____   3 × 2 = _____   3 × 1 = _____

3 × 3 = _____   3 × 1 = _____   3 × 4 = _____   3 × 1 = _____

3 × 5 = _____   3 × 1 = _____   3 × 2 = _____   3 × 3 = _____

3 × 2 = _____   3 × 4 = _____   3 × 2 = _____   3 × 5 = _____

3 × 2 = _____   3 × 1 = _____   3 × 2 = _____   3 × 3 = _____

3 × 1 = _____   3 × 3 = _____   3 × 2 = _____   3 × 3 = _____

3 × 4 = _____   3 × 3 = _____   3 × 5 = _____   3 × 3 = _____

3 × 4 = _____   3 × 1 = _____   3 × 4 = _____   3 × 2 = _____

3 × 4 = _____   3 × 3 = _____   3 × 4 = _____   3 × 5 = _____

3 × 4 = _____   3 × 5 = _____   3 × 1 = _____   3 × 5 = _____

3 × 2 = _____   3 × 5 = _____   3 × 3 = _____   3 × 5 = _____

3 × 4 = _____   3 × 2 = _____   3 × 4 = _____   3 × 3 = _____

3 × 5 = _____   3 × 3 = _____   3 × 2 = _____   3 × 4 = _____

3 × 3 = _____   3 × 5 = _____   3 × 2 = _____   3 × 4 = _____

multiply by 3 (1–5)

Lesson 1:   Solve word problems in varied contexts using a letter to represent the unknown.

© 2015 Great Minds®. eureka-math.org

Multiply.

$3 \times 1 =$ _____	$3 \times 2 =$ _____	$3 \times 3 =$ _____	$3 \times 4 =$ _____
$3 \times 5 =$ _____	$3 \times 6 =$ _____	$3 \times 7 =$ _____	$3 \times 8 =$ _____
$3 \times 9 =$ _____	$3 \times 10 =$ _____	$3 \times 5 =$ _____	$3 \times 6 =$ _____
$3 \times 5 =$ _____	$3 \times 7 =$ _____	$3 \times 5 =$ _____	$3 \times 8 =$ _____
$3 \times 5 =$ _____	$3 \times 9 =$ _____	$3 \times 5 =$ _____	$3 \times 10 =$ _____
$3 \times 6 =$ _____	$3 \times 5 =$ _____	$3 \times 6 =$ _____	$3 \times 7 =$ _____
$3 \times 6 =$ _____	$3 \times 8 =$ _____	$3 \times 6 =$ _____	$3 \times 9 =$ _____
$3 \times 6 =$ _____	$3 \times 7 =$ _____	$3 \times 6 =$ _____	$3 \times 7 =$ _____
$3 \times 8 =$ _____	$3 \times 7 =$ _____	$3 \times 9 =$ _____	$3 \times 7 =$ _____
$3 \times 8 =$ _____	$3 \times 6 =$ _____	$3 \times 8 =$ _____	$3 \times 7 =$ _____
$3 \times 8 =$ _____	$3 \times 9 =$ _____	$3 \times 9 =$ _____	$3 \times 6 =$ _____
$3 \times 9 =$ _____	$3 \times 7 =$ _____	$3 \times 9 =$ _____	$3 \times 8 =$ _____
$3 \times 9 =$ _____	$3 \times 8 =$ _____	$3 \times 6 =$ _____	$3 \times 9 =$ _____
$3 \times 7 =$ _____	$3 \times 9 =$ _____	$3 \times 6 =$ _____	$3 \times 8 =$ _____
$3 \times 9 =$ _____	$3 \times 7 =$ _____	$3 \times 6 =$ _____	$3 \times 8 =$ _____

multiply by 3 (6–10)

© 2015 Great Minds®. eureka-math.org

Multiply.

4 × 1 = _____	4 × 2 = _____	4 × 3 = _____	4 × 4 = _____
4 × 5 = _____	4 × 1 = _____	4 × 2 = _____	4 × 1 = _____
4 × 3 = _____	4 × 1 = _____	4 × 4 = _____	4 × 1 = _____
4 × 5 = _____	4 × 1 = _____	4 × 2 = _____	4 × 3 = _____
4 × 2 = _____	4 × 4 = _____	4 × 2 = _____	4 × 5 = _____
4 × 2 = _____	4 × 1 = _____	4 × 2 = _____	4 × 3 = _____
4 × 1 = _____	4 × 3 = _____	4 × 2 = _____	4 × 3 = _____
4 × 4 = _____	4 × 3 = _____	4 × 5 = _____	4 × 3 = _____
4 × 4 = _____	4 × 1 = _____	4 × 4 = _____	4 × 2 = _____
4 × 4 = _____	4 × 3 = _____	4 × 4 = _____	4 × 5 = _____
4 × 4 = _____	4 × 5 = _____	4 × 1 = _____	4 × 5 = _____
4 × 2 = _____	4 × 5 = _____	4 × 3 = _____	4 × 5 = _____
4 × 4 = _____	4 × 2 = _____	4 × 4 = _____	4 × 3 = _____
4 × 5 = _____	4 × 3 = _____	4 × 2 = _____	4 × 4 = _____
4 × 3 = _____	4 × 5 = _____	4 × 2 = _____	4 × 4 = _____

multiply by 4 (1–5)

Lesson 3:   Share and critique peer solution strategies to varied word problems.

103

© 2015 Great Minds®. eureka-math.org

Multiply.

4 × 1 = _____	4 × 2 = _____	4 × 3 = _____	4 × 4 = _____
4 × 5 = _____	4 × 6 = _____	4 × 7 = _____	4 × 8 = _____
4 × 9 = _____	4 × 10 = _____	4 × 5 = _____	4 × 6 = _____
4 × 5 = _____	4 × 7 = _____	4 × 5 = _____	4 × 8 = _____
4 × 5 = _____	4 × 9 = _____	4 × 5 = _____	4 × 10 = _____
4 × 6 = _____	4 × 5 = _____	4 × 6 = _____	4 × 7 = _____
4 × 6 = _____	4 × 8 = _____	4 × 6 = _____	4 × 9 = _____
4 × 6 = _____	4 × 7 = _____	4 × 6 = _____	4 × 7 = _____
4 × 8 = _____	4 × 7 = _____	4 × 9 = _____	4 × 7 = _____
4 × 8 = _____	4 × 6 = _____	4 × 8 = _____	4 × 7 = _____
4 × 8 = _____	4 × 9 = _____	4 × 9 = _____	4 × 6 = _____
4 × 9 = _____	4 × 7 = _____	4 × 9 = _____	4 × 8 = _____
4 × 9 = _____	4 × 8 = _____	4 × 6 = _____	4 × 9 = _____
4 × 7 = _____	4 × 9 = _____	4 × 6 = _____	4 × 8 = _____
4 × 9 = _____	4 × 7 = _____	4 × 6 = _____	4 × 8 = _____

multiply by 4 (6–10)

© 2015 Great Minds®. eureka-math.org

Multiply.

5 × 1 = _____   5 × 2 = _____   5 × 3 = _____   5 × 4 = _____

5 × 5 = _____   5 × 1 = _____   5 × 2 = _____   5 × 1 = _____

5 × 3 = _____   5 × 1 = _____   5 × 4 = _____   5 × 1 = _____

5 × 5 = _____   5 × 1 = _____   5 × 2 = _____   5 × 3 = _____

5 × 2 = _____   5 × 4 = _____   5 × 2 = _____   5 × 5 = _____

5 × 2 = _____   5 × 1 = _____   5 × 2 = _____   5 × 3 = _____

5 × 1 = _____   5 × 3 = _____   5 × 2 = _____   5 × 3 = _____

5 × 4 = _____   5 × 3 = _____   5 × 5 = _____   5 × 3 = _____

5 × 4 = _____   5 × 1 = _____   5 × 4 = _____   5 × 2 = _____

5 × 4 = _____   5 × 3 = _____   5 × 4 = _____   5 × 5 = _____

5 × 4 = _____   5 × 5 = _____   5 × 1 = _____   5 × 5 = _____

5 × 2 = _____   5 × 5 = _____   5 × 3 = _____   5 × 5 = _____

5 × 4 = _____   5 × 2 = _____   5 × 4 = _____   5 × 3 = _____

5 × 5 = _____   5 × 3 = _____   5 × 2 = _____   5 × 4 = _____

5 × 3 = _____   5 × 5 = _____   5 × 2 = _____   5 × 4 = _____

_____

multiply by 5 (1–5)

Lesson 5:   Compare and classify other polygons.

107

© 2015 Great Minds®. eureka-math.org

Multiply.

5 × 1 = _____   5 × 2 = _____   5 × 3 = _____   5 × 4 = _____

5 × 5 = _____   5 × 6 = _____   5 × 7 = _____   5 × 8 = _____

5 × 9 = _____   5 × 10 = _____   5 × 5 = _____   5 × 6 = _____

5 × 5 = _____   5 × 7 = _____   5 × 5 = _____   5 × 8 = _____

5 × 5 = _____   5 × 9 = _____   5 × 5 = _____   5 × 10 = _____

5 × 6 = _____   5 × 5 = _____   5 × 6 = _____   5 × 7 = _____

5 × 6 = _____   5 × 8 = _____   5 × 6 = _____   5 × 9 = _____

5 × 6 = _____   5 × 7 = _____   5 × 6 = _____   5 × 7 = _____

5 × 8 = _____   5 × 7 = _____   5 × 9 = _____   5 × 7 = _____

5 × 8 = _____   5 × 6 = _____   5 × 8 = _____   5 × 7 = _____

5 × 8 = _____   5 × 9 = _____   5 × 9 = _____   5 × 6 = _____

5 × 9 = _____   5 × 7 = _____   5 × 9 = _____   5 × 8 = _____

5 × 9 = _____   5 × 8 = _____   5 × 6 = _____   5 × 9 = _____

5 × 7 = _____   5 × 9 = _____   5 × 6 = _____   5 × 8 = _____

5 × 9 = _____   5 × 7 = _____   5 × 6 = _____   5 × 8 = _____

multiply by 5 (6–10)

Lesson 7:   Reason about composing and decomposing polygons using tetrominoes.

109

© 2015 Great Minds®. eureka-math.org

Multiply.

6 × 1 = _____   6 × 2 = _____   6 × 3 = _____   6 × 4 = _____

6 × 5 = _____   6 × 1 = _____   6 × 2 = _____   6 × 1 = _____

6 × 3 = _____   6 × 1 = _____   6 × 4 = _____   6 × 1 = _____

6 × 5 = _____   6 × 1 = _____   6 × 2 = _____   6 × 3 = _____

6 × 2 = _____   6 × 4 = _____   6 × 2 = _____   6 × 5 = _____

6 × 2 = _____   6 × 1 = _____   6 × 2 = _____   6 × 3 = _____

6 × 1 = _____   6 × 3 = _____   6 × 2 = _____   6 × 3 = _____

6 × 4 = _____   6 × 3 = _____   6 × 5 = _____   6 × 3 = _____

6 × 4 = _____   6 × 1 = _____   6 × 4 = _____   6 × 2 = _____

6 × 4 = _____   6 × 3 = _____   6 × 4 = _____   6 × 5 = _____

6 × 4 = _____   6 × 5 = _____   6 × 1 = _____   6 × 5 = _____

6 × 2 = _____   6 × 5 = _____   6 × 3 = _____   6 × 5 = _____

6 × 4 = _____   6 × 2 = _____   6 × 4 = _____   6 × 3 = _____

6 × 5 = _____   6 × 3 = _____   6 × 2 = _____   6 × 4 = _____

6 × 3 = _____   6 × 5 = _____   6 × 2 = _____   6 × 4 = _____

multiply by 6 (1–5)

Lesson 8:   Create a tangram puzzle and observe relationships among the shapes.

111

© 2015 Great Minds®. eureka-math.org

Multiply.

6 × 1 = _____	6 × 2 = _____	6 × 3 = _____	6 × 4 = _____
6 × 5 = _____	6 × 6 = _____	6 × 7 = _____	6 × 8 = _____
6 × 9 = _____	6 × 10 = _____	6 × 5 = _____	6 × 6 = _____
6 × 5 = _____	6 × 7 = _____	6 × 5 = _____	6 × 8 = _____
6 × 5 = _____	6 × 9 = _____	6 × 5 = _____	6 × 10 = _____
6 × 6 = _____	6 × 5 = _____	6 × 6 = _____	6 × 7 = _____
6 × 6 = _____	6 × 8 = _____	6 × 6 = _____	6 × 9 = _____
6 × 6 = _____	6 × 7 = _____	6 × 6 = _____	6 × 7 = _____
6 × 8 = _____	6 × 7 = _____	6 × 9 = _____	6 × 7 = _____
6 × 8 = _____	6 × 6 = _____	6 × 8 = _____	6 × 7 = _____
6 × 8 = _____	6 × 9 = _____	6 × 9 = _____	6 × 6 = _____
6 × 9 = _____	6 × 7 = _____	6 × 9 = _____	6 × 8 = _____
6 × 9 = _____	6 × 8 = _____	6 × 6 = _____	6 × 9 = _____
6 × 7 = _____	6 × 9 = _____	6 × 6 = _____	6 × 8 = _____
6 × 9 = _____	6 × 7 = _____	6 × 6 = _____	6 × 8 = _____

multiply by 6 (6–10)

EUREKA MATH   **Lesson 9:**   Reason about composing and decomposing polygons using tangrams.   **113**

© 2015 Great Minds®. eureka-math.org

Multiply

7 × 1 = _____	7 × 2 = _____	7 × 3 = _____	7 × 4 = _____
7 × 5 = _____	7 × 1 = _____	7 × 2 = _____	7 × 1 = _____
7 × 3 = _____	7 × 1 = _____	7 × 4 = _____	7 × 1 = _____
7 × 5 = _____	7 × 1 = _____	7 × 2 = _____	7 × 3 = _____
7 × 2 = _____	7 × 4 = _____	7 × 2 = _____	7 × 5 = _____
7 × 2 = _____	7 × 1 = _____	7 × 2 = _____	7 × 3 = _____
7 × 1 = _____	7 × 3 = _____	7 × 2 = _____	7 × 3 = _____
7 × 4 = _____	7 × 3 = _____	7 × 5 = _____	7 × 3 = _____
7 × 4 = _____	7 × 1 = _____	7 × 4 = _____	7 × 2 = _____
7 × 4 = _____	7 × 3 = _____	7 × 4 = _____	7 × 5 = _____
7 × 4 = _____	7 × 5 = _____	7 × 1 = _____	7 × 5 = _____
7 × 2 = _____	7 × 5 = _____	7 × 3 = _____	7 × 5 = _____
7 × 4 = _____	7 × 2 = _____	7 × 4 = _____	7 × 3 = _____
7 × 5 = _____	7 × 3 = _____	7 × 2 = _____	7 × 4 = _____
7 × 3 = _____	7 × 5 = _____	7 × 2 = _____	7 × 4 = _____

multiply by 7 (1–5)

Lesson 10:   Decompose quadrilaterals to understand perimeter as the boundary of a shape.

115

© 2015 Great Minds®. eureka-math.org

Multiply.

7 × 1 = _____	7 × 2 = _____	7 × 3 = _____	7 × 4 = _____
7 × 5 = _____	7 × 6 = _____	7 × 7 = _____	7 × 8 = _____
7 × 9 = _____	7 × 10 = _____	7 × 5 = _____	7 × 6 = _____
7 × 5 = _____	7 × 7 = _____	7 × 5 = _____	7 × 8 = _____
7 × 5 = _____	7 × 9 = _____	7 × 5 = _____	7 × 10 = _____
7 × 6 = _____	7 × 5 = _____	7 × 6 = _____	7 × 7 = _____
7 × 6 = _____	7 × 8 = _____	7 × 6 = _____	7 × 9 = _____
7 × 6 = _____	7 × 7 = _____	7 × 6 = _____	7 × 7 = _____
7 × 8 = _____	7 × 7 = _____	7 × 9 = _____	7 × 7 = _____
7 × 8 = _____	7 × 6 = _____	7 × 8 = _____	7 × 7 = _____
7 × 8 = _____	7 × 9 = _____	7 × 9 = _____	7 × 6 = _____
7 × 9 = _____	7 × 7 = _____	7 × 9 = _____	7 × 8 = _____
7 × 9 = _____	7 × 8 = _____	7 × 6 = _____	7 × 9 = _____
7 × 7 = _____	7 × 9 = _____	7 × 6 = _____	7 × 8 = _____
7 × 9 = _____	7 × 7 = _____	7 × 6 = _____	7 × 8 = _____

multiply by 7 (6–10)

Lesson 12: Measure side lengths in whole number units to determine the perimeter of polygons.

117

© 2015 Great Minds®. eureka-math.org

Multiply.

8 × 1 = _____	8 × 2 = _____	8 × 3 = _____	8 × 4 = _____
8 × 5 = _____	8 × 1 = _____	8 × 2 = _____	8 × 1 = _____
8 × 3 = _____	8 × 1 = _____	8 × 4 = _____	8 × 1 = _____
8 × 5 = _____	8 × 1 = _____	8 × 2 = _____	8 × 3 = _____
8 × 2 = _____	8 × 4 = _____	8 × 2 = _____	8 × 5 = _____
8 × 2 = _____	8 × 1 = _____	8 × 2 = _____	8 × 3 = _____
8 × 1 = _____	8 × 3 = _____	8 × 2 = _____	8 × 3 = _____
8 × 4 = _____	8 × 3 = _____	8 × 5 = _____	8 × 3 = _____
8 × 4 = _____	8 × 1 = _____	8 × 4 = _____	8 × 2 = _____
8 × 4 = _____	8 × 3 = _____	8 × 4 = _____	8 × 5 = _____
8 × 4 = _____	8 × 5 = _____	8 × 1 = _____	8 × 5 = _____
8 × 2 = _____	8 × 5 = _____	8 × 3 = _____	8 × 5 = _____
8 × 4 = _____	8 × 2 = _____	8 × 4 = _____	8 × 3 = _____
8 × 5 = _____	8 × 3 = _____	8 × 2 = _____	8 × 4 = _____
8 × 3 = _____	8 × 5 = _____	8 × 2 = _____	8 × 4 = _____

multiply by 8 (1–5)

© 2015 Great Minds®. eureka-math.org

Multiply.

8 × 1 = _____	8 × 2 = _____	8 × 3 = _____	8 × 4 = _____
8 × 5 = _____	8 × 6 = _____	8 × 7 = _____	8 × 8 = _____
8 × 9 = _____	8 × 10 = _____	8 × 5 = _____	8 × 6 = _____
8 × 5 = _____	8 × 7 = _____	8 × 5 = _____	8 × 8 = _____
8 × 5 = _____	8 × 9 = _____	8 × 5 = _____	8 × 10 = _____
8 × 6 = _____	8 × 5 = _____	8 × 6 = _____	8 × 7 = _____
8 × 6 = _____	8 × 8 = _____	8 × 6 = _____	8 × 9 = _____
8 × 6 = _____	8 × 7 = _____	8 × 6 = _____	8 × 7 = _____
8 × 8 = _____	8 × 7 = _____	8 × 9 = _____	8 × 7 = _____
8 × 8 = _____	8 × 6 = _____	8 × 8 = _____	8 × 7 = _____
8 × 8 = _____	8 × 9 = _____	8 × 9 = _____	8 × 6 = _____
8 × 9 = _____	8 × 7 = _____	8 × 9 = _____	8 × 8 = _____
8 × 9 = _____	8 × 8 = _____	8 × 6 = _____	8 × 9 = _____
8 × 7 = _____	8 × 9 = _____	8 × 6 = _____	8 × 8 = _____
8 × 9 = _____	8 × 7 = _____	8 × 6 = _____	8 × 8 = _____

multiply by 8 (6–10)

Lesson 14: Determine the perimeter of regular polygons and rectangles when whole number measurements are unknown.

© 2015 Great Minds®. eureka-math.org

Multiply.

9 × 1 = _____   9 × 2 = _____   9 × 3 = _____   9 × 4 = _____

9 × 5 = _____   9 × 1 = _____   9 × 2 = _____   9 × 1 = _____

9 × 3 = _____   9 × 1 = _____   9 × 4 = _____   9 × 1 = _____

9 × 5 = _____   9 × 1 = _____   9 × 2 = _____   9 × 3 = _____

9 × 2 = _____   9 × 4 = _____   9 × 2 = _____   9 × 5 = _____

9 × 2 = _____   9 × 1 = _____   9 × 2 = _____   9 × 3 = _____

9 × 1 = _____   9 × 3 = _____   9 × 2 = _____   9 × 3 = _____

9 × 4 = _____   9 × 3 = _____   9 × 5 = _____   9 × 3 = _____

9 × 4 = _____   9 × 1 = _____   9 × 4 = _____   9 × 2 = _____

9 × 4 = _____   9 × 3 = _____   9 × 4 = _____   9 × 5 = _____

9 × 4 = _____   9 × 5 = _____   9 × 1 = _____   9 × 5 = _____

9 × 2 = _____   9 × 5 = _____   9 × 3 = _____   9 × 5 = _____

9 × 4 = _____   9 × 2 = _____   9 × 4 = _____   9 × 3 = _____

9 × 5 = _____   9 × 3 = _____   9 × 2 = _____   9 × 4 = _____

9 × 3 = _____   9 × 5 = _____   9 × 2 = _____   9 × 4 = _____

multiply by 9 (1–5)

Lesson 15:   Solve word problems to determine perimeter with given side lengths.   **123**

© 2015 Great Minds®. eureka-math.org

Multiply.

9 × 1 = _____	9 × 2 = _____	9 × 3 = _____	9 × 4 = _____
9 × 5 = _____	9 × 6 = _____	9 × 7 = _____	9 × 8 = _____
9 × 9 = _____	9 × 10 = _____	9 × 5 = _____	9 × 6 = _____
9 × 5 = _____	9 × 7 = _____	9 × 5 = _____	9 × 8 = _____
9 × 5 = _____	9 × 9 = _____	9 × 5 = _____	9 × 10 = _____
9 × 6 = _____	9 × 5 = _____	9 × 6 = _____	9 × 7 = _____
9 × 6 = _____	9 × 8 = _____	9 × 6 = _____	9 × 9 = _____
9 × 6 = _____	9 × 7 = _____	9 × 6 = _____	9 × 7 = _____
9 × 8 = _____	9 × 7 = _____	9 × 9 = _____	9 × 7 = _____
9 × 8 = _____	9 × 6 = _____	9 × 8 = _____	9 × 7 = _____
9 × 8 = _____	9 × 9 = _____	9 × 9 = _____	9 × 6 = _____
9 × 9 = _____	9 × 7 = _____	9 × 9 = _____	9 × 8 = _____
9 × 9 = _____	9 × 8 = _____	9 × 6 = _____	9 × 9 = _____
9 × 7 = _____	9 × 9 = _____	9 × 6 = _____	9 × 8 = _____
9 × 9 = _____	9 × 7 = _____	9 × 6 = _____	9 × 8 = _____

multiply by 9 (6–10)

Lesson 16:   Use string to measure the perimeter of various circles to the nearest quarter inch.

© 2015 Great Minds®. eureka-math.org

A

Number Correct: _____

Multiply or Divide by 2

1.	$2 \times 2 =$		23.	___ $\times 2 = 20$	
2.	$3 \times 2 =$		24.	___ $\times 2 = 4$	
3.	$4 \times 2 =$		25.	___ $\times 2 = 6$	
4.	$5 \times 2 =$		26.	$20 \div 2 =$	
5.	$1 \times 2 =$		27.	$10 \div 2 =$	
6.	$4 \div 2 =$		28.	$2 \div 1 =$	
7.	$6 \div 2 =$		29.	$4 \div 2 =$	
8.	$10 \div 2 =$		30.	$6 \div 2 =$	
9.	$2 \div 1 =$		31.	___ $\times 2 = 12$	
10.	$8 \div 2 =$		32.	___ $\times 2 = 14$	
11.	$6 \times 2 =$		33.	___ $\times 2 = 18$	
12.	$7 \times 2 =$		34.	___ $\times 2 = 16$	
13.	$8 \times 2 =$		35.	$14 \div 2 =$	
14.	$9 \times 2 =$		36.	$18 \div 2 =$	
15.	$10 \times 2 =$		37.	$12 \div 2 =$	
16.	$16 \div 2 =$		38.	$16 \div 2 =$	
17.	$14 \div 2 =$		39.	$11 \times 2 =$	
18.	$18 \div 2 =$		40.	$22 \div 2 =$	
19.	$12 \div 2 =$		41.	$12 \times 2 =$	
20.	$20 \div 2 =$		42.	$24 \div 2 =$	
21.	___ $\times 2 = 10$		43.	$14 \times 2 =$	
22.	___ $\times 2 = 12$		44.	$28 \div 2 =$	

Lesson 20: Construct rectangles with a given perimeter using unit squares and determine their areas.

127

© 2015 Great Minds®. eureka-math.org

B

Number Correct: _____

Multiply or Divide by 2

Improvement: _____

1.	1 × 2 =	
2.	2 × 2 =	
3.	3 × 2 =	
4.	4 × 2 =	
5.	5 × 2 =	
6.	6 ÷ 2 =	
7.	4 ÷ 2 =	
8.	8 ÷ 2 =	
9.	2 ÷ 1 =	
10.	10 ÷ 2 =	
11.	10 × 2 =	
12.	6 × 2 =	
13.	7 × 2 =	
14.	8 × 2 =	
15.	9 × 2 =	
16.	14 ÷ 2 =	
17.	12 ÷ 2 =	
18.	16 ÷ 2 =	
19.	20 ÷ 2 =	
20.	18 ÷ 2 =	
21.	___ × 2 = 12	
22.	___ × 2 = 10	

23.	___ × 2 = 4	
24.	___ × 2 = 20	
25.	___ × 2 = 6	
26.	4 ÷ 2 =	
27.	2 ÷ 1 =	
28.	20 ÷ 2 =	
29.	10 ÷ 2 =	
30.	6 ÷ 2 =	
31.	___ × 2 = 12	
32.	___ × 2 = 16	
33.	___ × 2 = 18	
34.	___ × 2 = 14	
35.	16 ÷ 2 =	
36.	18 ÷ 2 =	
37.	12 ÷ 2 =	
38.	14 ÷ 2 =	
39.	11 × 2 =	
40.	22 ÷ 2 =	
41.	12 × 2 =	
42.	24 ÷ 2 =	
43.	13 × 2 =	
44.	26 ÷ 2 =	

Lesson 20: Construct rectangles with a given perimeter using unit squares and determine their areas.

129

© 2015 Great Minds®. eureka-math.org

A

Number Correct: _____

Multiply or Divide by 3

1.	2 × 3 =		23.	___ × 3 = 30	
2.	3 × 3 =		24.	___ × 3 = 6	
3.	4 × 3 =		25.	___ × 3 = 9	
4.	5 × 3 =		26.	30 ÷ 3 =	
5.	1 × 3 =		27.	15 ÷ 3 =	
6.	6 ÷ 3 =		28.	3 ÷ 3 =	
7.	9 ÷ 3 =		29.	6 ÷ 3 =	
8.	15 ÷ 3 =		30.	9 ÷ 3 =	
9.	3 ÷ 3 =		31.	___ × 3 = 18	
10.	12 ÷ 3 =		32.	___ × 3 = 21	
11.	6 × 3 =		33.	___ × 3 = 27	
12.	7 × 3 =		34.	___ × 3 = 24	
13.	8 × 3 =		35.	21 ÷ 3 =	
14.	9 × 3 =		36.	27 ÷ 3 =	
15.	10 × 3 =		37.	18 ÷ 3 =	
16.	24 ÷ 3 =		38.	24 ÷ 3 =	
17.	21 ÷ 3 =		39.	11 × 3 =	
18.	27 ÷ 3 =		40.	33 ÷ 3 =	
19.	18 ÷ 3 =		41.	12 × 3 =	
20.	30 ÷ 3 =		42.	36 ÷ 3 =	
21.	___ × 3 = 15		43.	13 × 3 =	
22.	___ × 3 = 3		44.	39 ÷ 3 =	

Lesson 21:   Construct rectangles with a given perimeter using unit squares and determine their areas.

131

© 2015 Great Minds®. eureka-math.org

B

Number Correct: _____

Multiply or Divide by 3

Improvement: _____

1.	1 × 3 =	
2.	2 × 3 =	
3.	3 × 3 =	
4.	4 × 3 =	
5.	5 × 3 =	
6.	9 ÷ 3 =	
7.	6 ÷ 3 =	
8.	12 ÷ 3 =	
9.	3 ÷ 3 =	
10.	15 ÷ 3 =	
11.	10 × 3 =	
12.	6 × 3 =	
13.	7 × 3 =	
14.	8 × 3 =	
15.	9 × 3 =	
16.	21 ÷ 3 =	
17.	18 ÷ 3 =	
18.	24 ÷ 3 =	
19.	30 ÷ 3 =	
20.	27 ÷ 3 =	
21.	___ × 3 = 3	
22.	___ × 3 = 15	

23.	___ × 3 = 6	
24.	___ × 3 = 30	
25.	___ × 3 = 9	
26.	6 ÷ 3 =	
27.	3 ÷ 3 =	
28.	30 ÷ 3 =	
29.	15 ÷ 3 =	
30.	9 ÷ 3 =	
31.	___ × 3 = 18	
32.	___ × 3 = 24	
33.	___ × 3 = 27	
34.	___ × 3 = 21	
35.	24 ÷ 3 =	
36.	27 ÷ 3 =	
37.	18 ÷ 3 =	
38.	21 ÷ 3 =	
39.	11 × 3 =	
40.	33 ÷ 3 =	
41.	12 × 3 =	
42.	36 ÷ 3 =	
43.	13 × 3 =	
44.	39 ÷ 3 =	

Lesson 21: Construct rectangles with a given perimeter using unit squares and determine their areas.

133

© 2015 Great Minds®. eureka-math.org

A

Number Correct: _____

Multiply or Divide by 4

1.	2 × 4 =	
2.	3 × 4 =	
3.	4 × 4 =	
4.	5 × 4 =	
5.	1 × 4 =	
6.	8 ÷ 4 =	
7.	12 ÷ 4 =	
8.	20 ÷ 4 =	
9.	4 ÷ 4 =	
10.	16 ÷ 4 =	
11.	6 × 4 =	
12.	7 × 4 =	
13.	8 × 4 =	
14.	9 × 4 =	
15.	10 × 4 =	
16.	32 ÷ 4 =	
17.	28 ÷ 4 =	
18.	36 ÷ 4 =	
19.	24 ÷ 4 =	
20.	40 ÷ 4 =	
21.	___ × 4 = 20	
22.	___ × 4 = 4	

23.	___ × 4 = 40	
24.	___ × 4 = 8	
25.	___ × 4 = 12	
26.	40 ÷ 4 =	
27.	20 ÷ 4 =	
28.	4 ÷ 4 =	
29.	8 ÷ 4 =	
30.	12 ÷ 4 =	
31.	___ × 4 = 24	
32.	___ × 4 = 28	
33.	___ × 4 = 36	
34.	___ × 4 = 32	
35.	28 ÷ 4 =	
36.	36 ÷ 4 =	
37.	24 ÷ 4 =	
38.	32 ÷ 4 =	
39.	11 × 4 =	
40.	44 ÷ 4 =	
41.	12 ÷ 4 =	
42.	48 ÷ 4 =	
43.	14 × 4 =	
44.	56 ÷ 4 =	

Lesson 22: Use a line plot to record the number of rectangles constructed in Lessons 20 and 21.

© 2015 Great Minds®. eureka-math.org

B

Number Correct: _____

Multiply or Divide by 4

Improvement: _____

1.	1 × 4 =	
2.	2 × 4 =	
3.	3 × 4 =	
4.	4 × 4 =	
5.	5 × 4 =	
6.	12 ÷ 4 =	
7.	8 ÷ 4 =	
8.	16 ÷ 4 =	
9.	4 ÷ 4 =	
10.	20 ÷ 4 =	
11.	10 × 4 =	
12.	6 × 4 =	
13.	7 × 4 =	
14.	8 × 4 =	
15.	9 × 4 =	
16.	28 ÷ 4 =	
17.	24 ÷ 4 =	
18.	32 ÷ 4 =	
19.	40 ÷ 4 =	
20.	36 ÷ 4 =	
21.	___ × 4 = 4	
22.	___ × 4 = 20	

23.	___ × 4 = 8	
24.	___ × 4 = 40	
25.	___ × 4 = 12	
26.	8 ÷ 4 =	
27.	4 ÷ 4 =	
28.	40 ÷ 4 =	
29.	20 ÷ 4 =	
30.	12 ÷ 4 =	
31.	___ × 4 = 12	
32.	___ × 4 = 16	
33.	___ × 4 = 36	
34.	___ × 4 = 28	
35.	32 ÷ 4 =	
36.	36 ÷ 4 =	
37.	24 ÷ 4 =	
38.	28 ÷ 4 =	
39.	11 × 4 =	
40.	44 ÷ 4 =	
41.	12 × 4 =	
42.	48 ÷ 4 =	
43.	13 × 4 =	
44.	52 ÷ 4 =	

Lesson 22: Use a line plot to record the number of rectangles constructed in Lessons 20 and 21.

137

© 2015 Great Minds®. eureka-math.org

A

Number Correct: _____

Multiply or Divide by 5

1.	$2 \times 5 =$		23.	___ $\times 5 = 50$	
2.	$3 \times 5 =$		24.	___ $\times 5 = 10$	
3.	$4 \times 5 =$		25.	___ $\times 5 = 15$	
4.	$5 \times 5 =$		26.	$50 \div 5 =$	
5.	$1 \times 5 =$		27.	$25 \div 5 =$	
6.	$10 \div 5 =$		28.	$5 \div 5 =$	
7.	$15 \div 5 =$		29.	$10 \div 5 =$	
8.	$25 \div 5 =$		30.	$15 \div 5 =$	
9.	$5 \div 5 =$		31.	___ $\times 5 = 30$	
10.	$20 \div 5 =$		32.	___ $\times 5 = 35$	
11.	$6 \times 5 =$		33.	___ $\times 5 = 45$	
12.	$7 \times 5 =$		34.	___ $\times 5 = 40$	
13.	$8 \times 5 =$		35.	$35 \div 5 =$	
14.	$9 \times 5 =$		36.	$45 \div 5 =$	
15.	$10 \times 5 =$		37.	$30 \div 5 =$	
16.	$40 \div 5 =$		38.	$40 \div 5 =$	
17.	$35 \div 5 =$		39.	$11 \times 5 =$	
18.	$45 \div 5 =$		40.	$55 \div 5 =$	
19.	$30 \div 5 =$		41.	$15 \div 5 =$	
20.	$50 \div 5 =$		42.	$60 \div 5 =$	
21.	___ $\times 5 = 25$		43.	$12 \times 5 =$	
22.	___ $\times 5 = 5$		44.	$70 \div 5 =$	

© 2015 Great Minds®. eureka-math.org

B

Number Correct: _____

Multiply or Divide by 5

Improvement: _____

1.	1 × 5 =	
2.	2 × 5 =	
3.	3 × 5 =	
4.	4 × 5 =	
5.	5 × 5 =	
6.	15 ÷ 5 =	
7.	10 ÷ 5 =	
8.	20 ÷ 5 =	
9.	5 ÷ 5 =	
10.	25 ÷ 5 =	
11.	10 × 5 =	
12.	6 × 5 =	
13.	7 × 5 =	
14.	8 × 5 =	
15.	9 × 5 =	
16.	35 ÷ 5 =	
17.	30 ÷ 5 =	
18.	40 ÷ 5 =	
19.	50 ÷ 5 =	
20.	45 ÷ 5 =	
21.	___ × 5 = 5	
22.	___ × 5 = 25	

23.	___ × 5 = 10	
24.	___ × 5 = 50	
25.	___ × 5 = 15	
26.	10 ÷ 5 =	
27.	5 ÷ 5 =	
28.	50 ÷ 5 =	
29.	25 ÷ 5 =	
30.	15 ÷ 5 =	
31.	___ × 5 = 15	
32.	___ × 5 = 20	
33.	___ × 5 = 45	
34.	___ × 5 = 35	
35.	40 ÷ 5 =	
36.	45 ÷ 5 =	
37.	30 ÷ 5 =	
38.	35 ÷ 5 =	
39.	11 × 5 =	
40.	55 ÷ 5 =	
41.	12 × 5 =	
42.	60 ÷ 5 =	
43.	13 × 5 =	
44.	65 ÷ 5 =	

© 2015 Great Minds®. eureka-math.org

Multiply.

6 × 1 = _____	6 × 2 = _____	6 × 3 = _____	6 × 4 = _____
6 × 5 = _____	6 × 6 = _____	6 × 7 = _____	6 × 8 = _____
6 × 9 = _____	6 × 10 = _____	6 × 5 = _____	6 × 6 = _____
6 × 5 = _____	6 × 7 = _____	6 × 5 = _____	6 × 8 = _____
6 × 5 = _____	6 × 9 = _____	6 × 5 = _____	6 × 10 = _____
6 × 6 = _____	6 × 5 = _____	6 × 6 = _____	6 × 7 = _____
6 × 6 = _____	6 × 8 = _____	6 × 6 = _____	6 × 9 = _____
6 × 6 = _____	6 × 7 = _____	6 × 6 = _____	6 × 7 = _____
6 × 8 = _____	6 × 7 = _____	6 × 9 = _____	6 × 7 = _____
6 × 8 = _____	6 × 6 = _____	6 × 8 = _____	6 × 7 = _____
6 × 8 = _____	6 × 9 = _____	6 × 9 = _____	6 × 6 = _____
6 × 9 = _____	6 × 7 = _____	6 × 9 = _____	6 × 8 = _____
6 × 9 = _____	6 × 8 = _____	6 × 6 = _____	6 × 9 = _____
6 × 7 = _____	6 × 9 = _____	6 × 6 = _____	6 × 8 = _____
6 × 9 = _____	6 × 7 = _____	6 × 6 = _____	6 × 8 = _____

multiply by 6 (6–10)

Lesson 24: Use rectangles to draw a robot with specified perimeter measurements, and reason about the different areas that may be produced.

© 2015 Great Minds®. eureka-math.org

A

Multiply or Divide by 6

1.	$2 \times 6 =$	
2.	$3 \times 6 =$	
3.	$4 \times 6 =$	
4.	$5 \times 6 =$	
5.	$1 \times 6 =$	
6.	$12 \div 6 =$	
7.	$18 \div 6 =$	
8.	$30 \div 6 =$	
9.	$6 \div 6 =$	
10.	$24 \div 6 =$	
11.	$6 \times 6 =$	
12.	$7 \times 6 =$	
13.	$8 \times 6 =$	
14.	$9 \times 6 =$	
15.	$10 \times 6 =$	
16.	$48 \div 6 =$	
17.	$42 \div 6 =$	
18.	$54 \div 6 =$	
19.	$36 \div 6 =$	
20.	$60 \div 6 =$	
21.	___ $\times 6 = 30$	
22.	___ $\times 6 = 6$	

23.	___ $\times 6 = 60$	
24.	___ $\times 6 = 12$	
25.	___ $\times 6 = 18$	
26.	$60 \div 6 =$	
27.	$30 \div 6 =$	
28.	$6 \div 6 =$	
29.	$12 \div 6 =$	
30.	$18 \div 6 =$	
31.	___ $\times 6 = 36$	
32.	___ $\times 6 = 42$	
33.	___ $\times 6 = 54$	
34.	___ $\times 6 = 48$	
35.	$42 \div 6 =$	
36.	$54 \div 6 =$	
37.	$36 \div 6 =$	
38.	$48 \div 6 =$	
39.	$11 \times 6 =$	
40.	$66 \div 6 =$	
41.	$12 \times 6 =$	
42.	$72 \div 6 =$	
43.	$14 \times 6 =$	
44.	$84 \div 6 =$	

Lesson 25:   Use rectangles to draw a robot with specified perimeter measurements, and reason about the different areas that may be produced.

© 2015 Great Minds®. eureka-math.org

145

B

Number Correct: _____

Multiply or Divide by 6

Improvement: _____

1.	1 × 6 =	
2.	2 × 6 =	
3.	3 × 6 =	
4.	4 × 6 =	
5.	5 × 6 =	
6.	18 ÷ 6 =	
7.	12 ÷ 6 =	
8.	24 ÷ 6 =	
9.	6 ÷ 6 =	
10.	30 ÷ 6 =	
11.	10 × 6 =	
12.	6 × 6 =	
13.	7 × 6 =	
14.	8 × 6 =	
15.	9 × 6 =	
16.	42 ÷ 6 =	
17.	36 ÷ 6 =	
18.	48 ÷ 6 =	
19.	60 ÷ 6 =	
20.	54 ÷ 6 =	
21.	___ × 6 = 6	
22.	___ × 6 = 30	

23.	___ × 6 = 12	
24.	___ × 6 = 60	
25.	___ × 6 = 18	
26.	12 ÷ 6 =	
27.	6 ÷ 6 =	
28.	60 ÷ 6 =	
29.	30 ÷ 6 =	
30.	18 ÷ 6 =	
31.	___ × 6 = 18	
32.	___ × 6 = 24	
33.	___ × 6 = 54	
34.	___ × 6 = 42	
35.	48 ÷ 6 =	
36.	54 ÷ 6 =	
37.	36 ÷ 6 =	
38.	42 ÷ 6 =	
39.	11 × 6 =	
40.	66 ÷ 6 =	
41.	12 × 6 =	
42.	72 ÷ 6 =	
43.	13 × 6 =	
44.	78 ÷ 6 =	

EUREKA MATH®

Lesson 25: Use rectangles to draw a robot with specified perimeter measurements, and reason about the different areas that may be produced.

147

© 2015 Great Minds®. eureka-math.org

Multiply.

7 × 1 = _____	7 × 2 = _____	7 × 3 = _____	7 × 4 = _____
7 × 5 = _____	7 × 6 = _____	7 × 7 = _____	7 × 8 = _____
7 × 9 = _____	7 × 10 = _____	7 × 5 = _____	7 × 6 = _____
7 × 5 = _____	7 × 7 = _____	7 × 5 = _____	7 × 8 = _____
7 × 5 = _____	7 × 9 = _____	7 × 5 = _____	7 × 10 = _____
7 × 6 = _____	7 × 5 = _____	7 × 6 = _____	7 × 7 = _____
7 × 6 = _____	7 × 8 = _____	7 × 6 = _____	7 × 9 = _____
7 × 6 = _____	7 × 7 = _____	7 × 6 = _____	7 × 7 = _____
7 × 8 = _____	7 × 7 = _____	7 × 9 = _____	7 × 7 = _____
7 × 8 = _____	7 × 6 = _____	7 × 8 = _____	7 × 7 = _____
7 × 8 = _____	7 × 9 = _____	7 × 9 = _____	7 × 6 = _____
7 × 9 = _____	7 × 7 = _____	7 × 9 = _____	7 × 8 = _____
7 × 9 = _____	7 × 8 = _____	7 × 6 = _____	7 × 9 = _____
7 × 7 = _____	7 × 9 = _____	7 × 6 = _____	7 × 8 = _____
7 × 9 = _____	7 × 7 = _____	7 × 6 = _____	7 × 8 = _____

multiply by 7 (6–10)

Lesson 26: Use rectangles to draw a robot with specified perimeter measurements, and reason about the different areas that may be produced.

© 2015 Great Minds®. eureka-math.org

A

Number Correct: _____

Multiply or Divide by 7

1.	2 × 7 =	
2.	3 × 7 =	
3.	4 × 7 =	
4.	5 × 7 =	
5.	1 × 7 =	
6.	14 ÷ 7 =	
7.	21 ÷ 7 =	
8.	35 ÷ 7 =	
9.	7 ÷ 7 =	
10.	28 ÷ 7 =	
11.	6 × 7 =	
12.	7 × 7 =	
13.	8 × 7 =	
14.	9 × 7 =	
15.	10 × 7 =	
16.	56 ÷ 7 =	
17.	49 ÷ 7 =	
18.	63 ÷ 7 =	
19.	42 ÷ 7 =	
20.	70 ÷ 7 =	
21.	___ × 7 = 35	
22.	___ × 7 = 7	

23.	___ × 7 = 70	
24.	___ × 7 = 14	
25.	___ × 7 = 21	
26.	70 ÷ 7 =	
27.	35 ÷ 7 =	
28.	7 ÷ 7 =	
29.	14 ÷ 7 =	
30.	21 ÷ 7 =	
31.	___ × 7 = 42	
32.	___ × 7 = 49	
33.	___ × 7 = 63	
34.	___ × 7 = 56	
35.	49 ÷ 7 =	
36.	63 ÷ 7 =	
37.	42 ÷ 7 =	
38.	56 ÷ 7 =	
39.	11 × 7 =	
40.	77 ÷ 7 =	
41.	12 × 7 =	
42.	84 ÷ 7 =	
43.	14 × 7 =	
44.	98 ÷ 7 =	

Lesson 27: Use rectangles to draw a robot with specified perimeter measurements, and reason about the different areas that may be produced.

151

© 2015 Great Minds®. eureka-math.org

B

Number Correct: _____

Multiply or Divide by 7

Improvement: _____

1.	1 × 7 =	
2.	2 × 7 =	
3.	3 × 7 =	
4.	4 × 7 =	
5.	5 × 7 =	
6.	21 ÷ 7 =	
7.	14 ÷ 7 =	
8.	28 ÷ 7 =	
9.	7 ÷ 7 =	
10.	35 ÷ 7 =	
11.	10 × 7 =	
12.	6 × 7 =	
13.	7 × 7 =	
14.	8 × 7 =	
15.	9 × 7 =	
16.	49 ÷ 7 =	
17.	42 ÷ 7 =	
18.	56 ÷ 7 =	
19.	70 ÷ 7 =	
20.	63 ÷ 7 =	
21.	___ × 7 = 7	
22.	___ × 7 = 35	

23.	___ × 7 = 14	
24.	___ × 7 = 70	
25.	___ × 7 = 21	
26.	14 ÷ 7 =	
27.	7 ÷ 7 =	
28.	70 ÷ 7 =	
29.	35 ÷ 7 =	
30.	21 ÷ 7 =	
31.	___ × 7 = 21	
32.	___ × 7 = 28	
33.	___ × 7 = 63	
34.	___ × 7 = 49	
35.	56 ÷ 7 =	
36.	63 ÷ 7 =	
37.	42 ÷ 7 =	
38.	49 ÷ 7 =	
39.	11 × 7 =	
40.	77 ÷ 7 =	
41.	12 × 7 =	
42.	84 ÷ 7 =	
43.	13 × 7 =	
44.	91 ÷ 7 =	

Lesson 27:   Use rectangles to draw a robot with specified perimeter measurements, and reason about the different areas that may be produced.

153

© 2015 Great Minds®. eureka-math.org

Multiply.

8 × 1 = _____	8 × 2 = _____	8 × 3 = _____	8 × 4 = _____
8 × 5 = _____	8 × 6 = _____	8 × 7 = _____	8 × 8 = _____
8 × 9 = _____	8 × 10 = _____	8 × 5 = _____	8 × 6 = _____
8 × 5 = _____	8 × 7 = _____	8 × 5 = _____	8 × 8 = _____
8 × 5 = _____	8 × 9 = _____	8 × 5 = _____	8 × 10 = _____
8 × 6 = _____	8 × 5 = _____	8 × 6 = _____	8 × 7 = _____
8 × 6 = _____	8 × 8 = _____	8 × 6 = _____	8 × 9 = _____
8 × 6 = _____	8 × 7 = _____	8 × 6 = _____	8 × 7 = _____
8 × 8 = _____	8 × 7 = _____	8 × 9 = _____	8 × 7 = _____
8 × 8 = _____	8 × 6 = _____	8 × 8 = _____	8 × 7 = _____
8 × 8 = _____	8 × 9 = _____	8 × 9 = _____	8 × 6 = _____
8 × 9 = _____	8 × 7 = _____	8 × 9 = _____	8 × 8 = _____
8 × 9 = _____	8 × 8 = _____	8 × 6 = _____	8 × 9 = _____
8 × 7 = _____	8 × 9 = _____	8 × 6 = _____	8 × 8 = _____
8 × 9 = _____	8 × 7 = _____	8 × 6 = _____	8 × 8 = _____

multiply by 8 (6–10)

Lesson 28: Solve a variety of word problems involving area and perimeter using all four operations.

© 2015 Great Minds®. eureka-math.org

A

Number Correct: _____

Multiply or Divide by 8

1.	2 × 8 =		23.	___ × 8 = 80	
2.	3 × 8 =		24.	___ × 8 = 16	
3.	4 × 8 =		25.	___ × 8 = 24	
4.	5 × 8 =		26.	80 ÷ 8 =	
5.	1 × 8 =		27.	40 ÷ 8 =	
6.	16 ÷ 8 =		28.	8 ÷ 8 =	
7.	24 ÷ 8 =		29.	16 ÷ 8 =	
8.	40 ÷ 8 =		30.	24 ÷ 8 =	
9.	8 ÷ 8 =		31.	___ × 8 = 48	
10.	32 ÷ 8 =		32.	___ × 8 = 56	
11.	6 × 8 =		33.	___ × 8 = 72	
12.	7 × 8 =		34.	___ × 8 = 64	
13.	8 × 8 =		35.	56 ÷ 8 =	
14.	9 × 8 =		36.	72 ÷ 8 =	
15.	10 × 8 =		37.	48 ÷ 8 =	
16.	64 ÷ 8 =		38.	64 ÷ 8 =	
17.	56 ÷ 8 =		39.	11 × 8 =	
18.	72 ÷ 8 =		40.	88 ÷ 8 =	
19.	48 ÷ 8 =		41.	12 × 8 =	
20.	80 ÷ 8 =		42.	96 ÷ 8 =	
21.	___ × 8 = 40		43.	14 × 8 =	
22.	___ × 8 = 8		44.	112 ÷ 8 =	

Lesson 29: Solve a variety of word problems involving area and perimeter using all four operations.

157

© 2015 Great Minds®. eureka-math.org

B

Number Correct: _____

Multiply or Divide by 8

Improvement: _____

1.	1 × 8 =		23.	___ × 8 = 16	
2.	2 × 8 =		24.	___ × 8 = 80	
3.	3 × 8 =		25.	___ × 8 = 24	
4.	4 × 8 =		26.	16 ÷ 8 =	
5.	5 × 8 =		27.	8 ÷ 8 =	
6.	24 ÷ 8 =		28.	80 ÷ 8 =	
7.	16 ÷ 8 =		29.	40 ÷ 8 =	
8.	32 ÷ 8 =		30.	24 ÷ 8 =	
9.	8 ÷ 8 =		31.	___ × 8 = 24	
10.	40 ÷ 8 =		32.	___ × 8 = 32	
11.	10 × 8 =		33.	___ × 8 = 72	
12.	6 × 8 =		34.	___ × 8 = 56	
13.	7 × 8 =		35.	64 ÷ 8 =	
14.	8 × 8 =		36.	72 ÷ 8 =	
15.	9 × 8 =		37.	48 ÷ 8 =	
16.	56 ÷ 8 =		38.	56 ÷ 8 =	
17.	8 ÷ 8 =		39.	11 × 8 =	
18.	64 ÷ 8 =		40.	88 ÷ 8 =	
19.	80 ÷ 8 =		41.	12 × 8 =	
20.	72 ÷ 8 =		42.	96 ÷ 8 =	
21.	___ × 8 = 8		43.	13 × 8 =	
22.	___ × 8 = 40		44.	104 ÷ 8 =	

Lesson 29: Solve a variety of word problems involving area and perimeter using all four operations.

159

© 2015 Great Minds®. eureka-math.org

Multiply.

9 × 1 = _____   9 × 2 = _____   9 × 3 = _____   9 × 4 = _____

9 × 5 = _____   9 × 6 = _____   9 × 7 = _____   9 × 8 = _____

9 × 9 = _____   9 × 10 = _____   9 × 5 = _____   9 × 6 = _____

9 × 5 = _____   9 × 7 = _____   9 × 5 = _____   9 × 8 = _____

9 × 5 = _____   9 × 9 = _____   9 × 5 = _____   9 × 10 = _____

9 × 6 = _____   9 × 5 = _____   9 × 6 = _____   9 × 7 = _____

9 × 6 = _____   9 × 8 = _____   9 × 6 = _____   9 × 9 = _____

9 × 6 = _____   9 × 7 = _____   9 × 6 = _____   9 × 7 = _____

9 × 8 = _____   9 × 7 = _____   9 × 9 = _____   9 × 7 = _____

9 × 8 = _____   9 × 6 = _____   9 × 8 = _____   9 × 7 = _____

9 × 8 = _____   9 × 9 = _____   9 × 9 = _____   9 × 6 = _____

9 × 9 = _____   9 × 7 = _____   9 × 9 = _____   9 × 8 = _____

9 × 9 = _____   9 × 8 = _____   9 × 6 = _____   9 × 9 = _____

9 × 7 = _____   9 × 9 = _____   9 × 6 = _____   9 × 8 = _____

9 × 9 = _____   9 × 7 = _____   9 × 6 = _____   9 × 8 = _____

multiply by 9 (6–10)

Lesson 30:   Share and critique peer strategies for problem solving.

161

© 2015 Great Minds®. eureka-math.org

A

Number Correct: _____

Multiply or Divide by 9

1.	2 × 9 =	
2.	3 × 9 =	
3.	4 × 9 =	
4.	5 × 9 =	
5.	1 × 9 =	
6.	18 ÷ 9 =	
7.	27 ÷ 9 =	
8.	45 ÷ 9 =	
9.	9 ÷ 9 =	
10.	36 ÷ 9 =	
11.	6 × 9 =	
12.	7 × 9 =	
13.	8 × 9 =	
14.	9 × 9 =	
15.	10 × 9 =	
16.	72 ÷ 9 =	
17.	63 ÷ 9 =	
18.	81 ÷ 9 =	
19.	54 ÷ 9 =	
20.	90 ÷ 9 =	
21.	___ × 9 = 45	
22.	___ × 9 = 9	

23.	___ × 9 = 90	
24.	___ × 9 = 18	
25.	___ × 9 = 27	
26.	90 ÷ 9 =	
27.	45 ÷ 9 =	
28.	9 ÷ 9 =	
29.	18 ÷ 9 =	
30.	27 ÷ 9 =	
31.	___ × 9 = 54	
32.	___ × 9 = 63	
33.	___ × 9 = 81	
34.	___ × 9 = 72	
35.	63 ÷ 9 =	
36.	81 ÷ 9 =	
37.	54 ÷ 9 =	
38.	72 ÷ 9 =	
39.	11 × 9 =	
40.	99 ÷ 9 =	
41.	12 × 9 =	
42.	108 ÷ 9 =	
43.	14 × 9 =	
44.	126 ÷ 9 =	

Lesson 31:   Explore and create unconventional representations of one-half.

163

© 2015 Great Minds®. eureka-math.org

B

Number Correct: _____

Multiply or Divide by 9

Improvement: _____

1.	1 × 9 =	
2.	2 × 9 =	
3.	3 × 9 =	
4.	4 × 9 =	
5.	5 × 9 =	
6.	27 ÷ 9 =	
7.	18 ÷ 9 =	
8.	36 ÷ 9 =	
9.	9 ÷ 9 =	
10.	45 ÷ 9 =	
11.	10 × 9 =	
12.	6 × 9 =	
13.	7 × 9 =	
14.	8 × 9 =	
15.	9 × 9 =	
16.	63 ÷ 9 =	
17.	54 ÷ 9 =	
18.	72 ÷ 9 =	
19.	90 ÷ 9 =	
20.	81 ÷ 9 =	
21.	___ × 9 = 9	
22.	___ × 9 = 45	

23.	___ × 9 = 18	
24.	___ × 9 = 90	
25.	___ × 9 = 27	
26.	18 ÷ 9 =	
27.	9 ÷ 9 =	
28.	90 ÷ 9 =	
29.	45 ÷ 9 =	
30.	27 ÷ 9 =	
31.	___ × 9 = 27	
32.	___ × 9 = 36	
33.	___ × 9 = 81	
34.	___ × 9 = 63	
35.	72 ÷ 9 =	
36.	81 ÷ 9 =	
37.	54 ÷ 9 =	
38.	63 ÷ 9 =	
39.	11 × 9 =	
40.	99 ÷ 9 =	
41.	12 × 9 =	
42.	108 ÷ 9 =	
43.	13 × 9 =	
44.	117 ÷ 9 =	

EUREKA MATH®

Lesson 31: Explore and create unconventional representations of one-half.

165

© 2015 Great Minds®. eureka-math.org

A

Number Correct: _____

Mixed Multiplication

1.	2 × 1 =		23.	2 × 7 =	
2.	2 × 2 =		24.	5 × 5 =	
3.	2 × 3 =		25.	5 × 6 =	
4.	4 × 1 =		26.	5 × 7 =	
5.	4 × 2 =		27.	4 × 5 =	
6.	4 × 3 =		28.	4 × 6 =	
7.	1 × 6 =		29.	4 × 7 =	
8.	2 × 6 =		30.	3 × 5 =	
9.	1 × 8 =		31.	3 × 6 =	
10.	2 × 8 =		32.	3 × 7 =	
11.	3 × 1 =		33.	2 × 7 =	
12.	3 × 2 =		34.	2 × 8 =	
13.	3 × 3 =		35.	2 × 9 =	
14.	5 × 1 =		36.	5 × 7 =	
15.	5 × 2 =		37.	5 × 8 =	
16.	5 × 3 =		38.	5 × 9 =	
17.	1 × 7 =		39.	4 × 7 =	
18.	2 × 7 =		40.	4 × 8 =	
19.	1 × 9 =		41.	4 × 9 =	
20.	2 × 9 =		42.	3 × 7 =	
21.	2 × 5 =		43.	3 × 8 =	
22.	2 × 6 =		44.	3 × 9 =	

Lesson 32: Explore and create unconventional representations of one-half.

© 2015 Great Minds®. eureka-math.org

B

Number Correct: _____

Mixed Multiplication

Improvement: _____

1.	5 × 1 =		23.	5 × 7 =	
2.	5 × 2 =		24.	2 × 5 =	
3.	5 × 3 =		25.	2 × 6 =	
4.	3 × 1 =		26.	2 × 7 =	
5.	3 × 2 =		27.	3 × 5 =	
6.	3 × 3 =		28.	3 × 6 =	
7.	1 × 7 =		29.	3 × 7 =	
8.	2 × 7 =		30.	4 × 5 =	
9.	1 × 9 =		31.	4 × 6 =	
10.	2 × 9 =		32.	4 × 7 =	
11.	2 × 1 =		33.	5 × 7 =	
12.	2 × 2 =		34.	5 × 8 =	
13.	2 × 3 =		35.	5 × 9 =	
14.	4 × 1 =		36.	2 × 7 =	
15.	4 × 2 =		37.	2 × 8 =	
16.	4 × 3 =		38.	2 × 9 =	
17.	1 × 6 =		39.	3 × 7 =	
18.	2 × 6 =		40.	3 × 8 =	
19.	1 × 8 =		41.	3 × 9 =	
20.	2 × 8 =		42.	4 × 7 =	
21.	5 × 5 =		43.	4 × 8 =	
22.	5 × 6 =		44.	4 × 9 =	

© 2015 Great Minds®. eureka-math.org

A

Number Correct: _____

Mixed Division

1.	4 ÷ 2 =		23.	16 ÷ 8 =	
2.	6 ÷ 2 =		24.	40 ÷ 8 =	
3.	10 ÷ 2 =		25.	32 ÷ 8 =	
4.	20 ÷ 2 =		26.	56 ÷ 8 =	
5.	10 ÷ 5 =		27.	18 ÷ 9 =	
6.	15 ÷ 5 =		28.	45 ÷ 9 =	
7.	25 ÷ 5 =		29.	36 ÷ 9 =	
8.	20 ÷ 5 =		30.	63 ÷ 9 =	
9.	8 ÷ 4 =		31.	64 ÷ 8 =	
10.	12 ÷ 4 =		32.	48 ÷ 8 =	
11.	20 ÷ 4 =		33.	81 ÷ 9 =	
12.	16 ÷ 4 =		34.	54 ÷ 9 =	
13.	6 ÷ 3 =		35.	24 ÷ 6 =	
14.	9 ÷ 3 =		36.	16 ÷ 2 =	
15.	15 ÷ 3 =		37.	28 ÷ 7 =	
16.	12 ÷ 3 =		38.	27 ÷ 3 =	
17.	60 ÷ 6 =		39.	24 ÷ 8 =	
18.	12 ÷ 6 =		40.	32 ÷ 4 =	
19.	18 ÷ 6 =		41.	27 ÷ 9 =	
20.	35 ÷ 7 =		42.	72 ÷ 9 =	
21.	14 ÷ 7 =		43.	56 ÷ 7 =	
22.	21 ÷ 7 =		44.	72 ÷ 8 =	

Lesson 33: Solidify fluency with Grade 3 skills.

© 2015 Great Minds®. eureka-math.org

B

Number Correct: _____

Improvement: _____

Mixed Division

1.	10 ÷ 5 =	
2.	15 ÷ 5 =	
3.	25 ÷ 5 =	
4.	50 ÷ 5 =	
5.	4 ÷ 2 =	
6.	6 ÷ 2 =	
7.	10 ÷ 2 =	
8.	8 ÷ 2 =	
9.	6 ÷ 3 =	
10.	9 ÷ 3 =	
11.	15 ÷ 3 =	
12.	12 ÷ 3 =	
13.	8 ÷ 4 =	
14.	12 ÷ 4 =	
15.	20 ÷ 4 =	
16.	16 ÷ 4 =	
17.	70 ÷ 7 =	
18.	14 ÷ 7 =	
19.	21 ÷ 7 =	
20.	30 ÷ 6 =	
21.	12 ÷ 6 =	
22.	18 ÷ 6 =	

23.	18 ÷ 9 =	
24.	45 ÷ 9 =	
25.	27 ÷ 9 =	
26.	63 ÷ 9 =	
27.	16 ÷ 8 =	
28.	40 ÷ 8 =	
29.	24 ÷ 8 =	
30.	56 ÷ 8 =	
31.	81 ÷ 9 =	
32.	54 ÷ 9 =	
33.	64 ÷ 8 =	
34.	48 ÷ 8 =	
35.	30 ÷ 6 =	
36.	18 ÷ 2 =	
37.	35 ÷ 7 =	
38.	24 ÷ 3 =	
39.	32 ÷ 8 =	
40.	36 ÷ 4 =	
41.	45 ÷ 9 =	
42.	72 ÷ 8 =	
43.	49 ÷ 7 =	
44.	72 ÷ 9 =	

Lesson 33:   Solidify fluency with Grade 3 skills.

© 2015 Great Minds®. eureka-math.org

A

Number Correct: _____

Multiply and Divide

1.	3 × 2 =	
2.	6 ÷ 2 =	
3.	5 × 3 =	
4.	15 ÷ 5 =	
5.	4 × 2 =	
6.	8 ÷ 4 =	
7.	3 × 3 =	
8.	9 ÷ 3 =	
9.	4 × 3 =	
10.	12 ÷ 4 =	
11.	5 × 5 =	
12.	25 ÷ 5 =	
13.	6 × 2 =	
14.	21 ÷ 7 =	
15.	7 × 4 =	
16.	16 ÷ 8 =	
17.	18 ÷ 3 =	
18.	18 ÷ 9 =	
19.	8 × 3 =	
20.	36 ÷ 9 =	
21.	14 ÷ 7 =	
22.	6 × 4 =	

23.	2 × 7 =	
24.	3 × 8 =	
25.	4 × 9 =	
26.	5 × 7 =	
27.	36 ÷ 6 =	
28.	42 ÷ 7 =	
29.	64 ÷ 8 =	
30.	45 ÷ 9 =	
31.	2 × 8 =	
32.	3 × 9 =	
33.	32 ÷ 4 =	
34.	45 ÷ 5 =	
35.	6 × 7 =	
36.	7 × 7 =	
37.	56 ÷ 8 =	
38.	63 ÷ 9 =	
39.	6 × 6 =	
40.	8 × 8 =	
41.	81 ÷ 9 =	
42.	49 ÷ 7 =	
43.	54 ÷ 6 =	
44.	56 ÷ 7 =	

EUREKA MATH

Lesson 34: Create resource booklets to support fluency with Grade 3 skills.

175

© 2015 Great Minds®. eureka-math.org

B

Number Correct: _____

Multiply and Divide

Improvement: _____

1.	$5 \times 2 =$	
2.	$10 \div 2 =$	
3.	$2 \times 3 =$	
4.	$6 \div 3 =$	
5.	$3 \times 2 =$	
6.	$6 \div 2 =$	
7.	$4 \times 4 =$	
8.	$16 \div 4 =$	
9.	$3 \times 4 =$	
10.	$12 \div 3 =$	
11.	$3 \times 3 =$	
12.	$9 \div 3 =$	
13.	$7 \times 2 =$	
14.	$18 \div 6 =$	
15.	$6 \times 4 =$	
16.	$18 \div 9 =$	
17.	$21 \div 3 =$	
18.	$16 \div 8 =$	
19.	$9 \times 3 =$	
20.	$32 \div 8 =$	
21.	$12 \div 6 =$	
22.	$7 \times 4 =$	

23.	$2 \times 7 =$	
24.	$3 \times 8 =$	
25.	$4 \times 9 =$	
26.	$5 \times 7 =$	
27.	$36 \div 6 =$	
28.	$42 \div 7 =$	
29.	$64 \div 8 =$	
30.	$45 \div 9 =$	
31.	$2 \times 8 =$	
32.	$3 \times 9 =$	
33.	$32 \div 4 =$	
34.	$45 \div 5 =$	
35.	$6 \times 7 =$	
36.	$7 \times 7 =$	
37.	$56 \div 8 =$	
38.	$63 \div 9 =$	
39.	$6 \times 6 =$	
40.	$8 \times 8 =$	
41.	$81 \div 9 =$	
42.	$49 \div 7 =$	
43.	$54 \div 6 =$	
44.	$56 \div 7 =$	

Lesson 34: Create resource booklets to support fluency with Grade 3 skills.

177

© 2015 Great Minds®. eureka-math.org

Credits

Great Minds® has made every effort to obtain permission for the reprinting of all copyrighted material. If any owner of copyrighted material is not acknowledged herein, please contact Great Minds for proper acknowledgment in all future editions and reprints of this module.